Intermediate Health & Safety

A text for Intermediate Health & Safety courses and a reference for supervisors

Ian Fisher
MCIEH FRSH FRIPH MIIRSM

PUBLISHED BY
© HIGHFIELD.CO.UK LTD

"Vue Pointe", Spinney Hill, Sprotbrough,
Doncaster, DN5 7LY, UK
Tel: +44 0845 2260350
Fax: +44 0845 2260360
E-mail: richard@highfield.co.uk

Websites:
www.highfield.co.uk
www.foodsafetytrainers.co.ul

ISBN 1 904544 42 8

Highfield .co.uk limited

This book is dedicated to those who have lost their life or suffered irreparable damage to their health whilst at work.

First published 2004
2nd Edition 2005

ISBN 1 904544 42 8

Printed by Trafford Press • Telephone: 01302 367509

Contents

Health
& Safety
at Work
Assembly
point

Foreword

The hazards and risks faced at work have changed for many of us over the years with the demographic changes in employment. It is also interesting to note that the perception of health and safety at work held by managers and supervisors has also changed.

More than 150 years ago it was recognised by society, famous authors and philanthropists that the work we do, together with where and how we do it, are important factors in our general health. Legal requirements were introduced and became the standard to be achieved.

Then, over the years, the potential financial losses of poor management of health and safety became more important to owners and managers and financial control drove up standards.

This is now changing again. We now live in a society where the fear of litigation and the requirement of insurers are beginning to drive employers' actions.

What has not changed is the difficulty of the role of the line manager or supervisor. They have to demand safe working practice whilst under pressure from their manager to get the job done and yet working themselves with friends and colleagues.

In recognition of the importance of the role of the supervisor in effective health and safety management, this book is not a detailed text book of health and safety legislation. It tries to explain the legal requirements and good management practice in a simple, common sense way. It has been written to support supervisors and managers that have health and safety as one of their responsibilities, and in particular as a text for any Level 2 health and safety qualification.

Ian Fisher
MCIEH FRSH FRIPH MIIRSM

Ian is an experienced Environmental Health Officer, working as a consultant for over a decade. He has specialised in Risk Management and training, particularly in the development of health and safety management systems for multiple outlet clients. In his not so spare time Ian works actively for his local community, the village of Waingroves and is a Non-Executive Director of an NHS Trust.

Acknowledgments

This book has been written to support those who manage or supervise staff and find themselves responsible for health and safety at work. My experience of working with and training those very people has fed the content. So thanks to everyone who has taught me so much – particularly all those I met working in the NHS, the healthcare business and for Marks and Spencer.

My thoughts have ended up on paper due to the combined efforts of the staff at Highfield, a remarkable bunch of friendly people who work to incredible, tight timetables and yet produce work of the highest standard. In particular thanks to Kirstie Atkinson for her pragmatic advice and checking the accuracy.

When working with a subject such as health and safety at work, we are exposed to many books and notes on the topic. Most of the detail is based on HSE publications, and I freely acknowledge them as a source of information we should all follow. However, there may be words that someone recognises as their own. For this I apologise and if you contact me they will be acknowledged in any future editions of this book.

I may have been tempted to use too many words, but thanks to my wife Dawn who has used her skills to proof read it to make sure it does not contain too much jargon.

Finally thanks to Richard Sprenger, who encouraged me to take the time to write the book.

Ian Fisher

1 Introduction

So often managers and supervisors explain why something has to be done as 'It's health and safety.'

The above statement makes this question interesting:

WHY IS HEALTH AND SAFETY AT WORK IMPORTANT?

There are three major reasons why health and safety at work is important to everyone:

A) Moral – Employers and their staff want to provide the highest possible standards of service provided for customers. An all important part of this, is a desire to see a safe and risk-free environment for all colleagues and visitors. It has to be remembered that an injury caused at work affects every aspect of life, not just the time spent at work.

B) Statutory - Legislation on health and safety at work has a great influence on most employers. Usually it is the owner, chief executive or directors who are exposed to the risk of criminal prosecution if they do not comply with the law. Maximum fines are usually in the area of £20,000 for each offence, but there is provision for larger fines or even a prison sentence in exceptional cases.

Although there have been many prosecutions against employers, there is a view that we should not just be motivated by statutory requirements, and that the legislation, approved codes of practice and HSE guidance give us a lead on how to implement effective health and safety at work.

C) Financial - There is now, more than ever before, a likelihood of individuals that have suffered loss taking civil action. Solicitors advertise for clients with messages such as 'Have you had an accident in the last three years?' For every claim there are many costs, including insurance and legal advice.

In addition every accident has great potential for financial cost to the employer from these items:

Costs of poor health and safety at work:

- compensation – including legal costs and increased insurance premiums;
- loss of production – if someone is not at work due to injury or illness, colleagues may have to cover, increasing stress and leading to more mistakes;
- extra staff – more employees are needed or agency staff may be used;
- treatment – first aid;
- cleaning up – not just literal cleaning up but repairs to buildings;
- prosecution – including legal costs; and
- reputation – it is difficult to gain a good reputation. It is easy to lose it when the media carry stories of serious incidents or prosecutions. Staff and customers lose confidence.

Benefits of good health and safety management

These are the opposite to the costs, but there is a clear link between good management and quality and consistency. If health and safety is managed well, with clearly defined policies and procedures, and well-trained and highly-motivated staff, usually the quality and consistency of the product or service will improve.

Line managers and supervisors

Everyone recognises that line managers and supervisors have a difficult job. They are usually responsible for making sure that policies and procedures set by management are implemented by everyone - and yet they often have a job to do themselves.

Line managers and supervisors have an essential role in effective management – the day-to-day control. If they display a positive attitude and enthusiasm, these will be staff, motivated to do a good job.

Supervisors also have specific tasks:

- training;
- inspection;
- measuring performance;
- consulting and involving others;
- assessing risks; and
- assisting managers.

However, there is one role of supervisors that is always difficult to carry out, that of disciplining the staff. There is a point when encouragement and training have to stop and staff have to be told what to do. If they persist in ignoring instructions, it is usually the line manager or supervisor that begins disciplinary action. This can lead to conflict in the workplace, but with the correct communication skills and common sense standards applied, health and safety does not have to become a battleground.

2 Accidents

In spite of legislation aimed at preventing accidents and considerable efforts made by employers to reduce costs, accidents continue to happen. An important part of accident prevention and the effective management of claims is local knowledge of the accidents that happen within our place of work. This requires robust policies and procedures that aim to provide compliance with legislation and to fulfil your needs for information.

The accident triangle

In a typical workplace, it has been shown that there is a relationship between near misses, minor accidents and major accidents. This can be illustrated by the accident triangle:

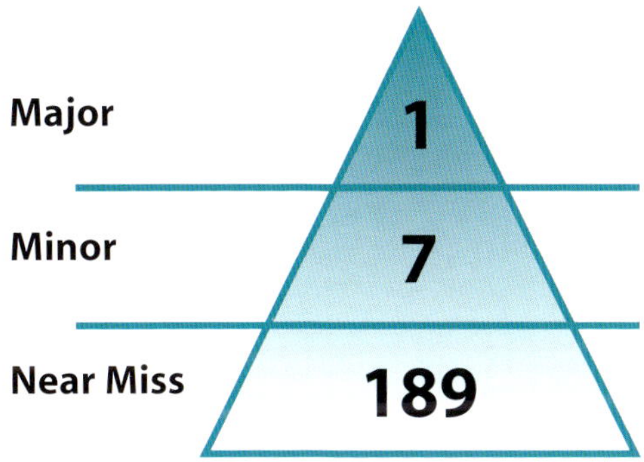

For every major, disabling injury, there are 7 less serious injuries, requiring first aid or time off work, and 189 near misses.

This shows the importance of reporting all incidents, as the action we take to prevent a recurrence of a near miss could prevent a very serious incident. This is important when staff are trained – they must report all incidents, no matter how trivial they think they are. Managers should then analyse the data to establish trends and take action to prevent the same accident happening again.

LEGISLATION

The following legislation applies to accident and incident reporting:

The Social Security Act and Regulations

These are concerned with the benefits paid to people unable to work, including injury at work. To assist the determination of the right to benefit under the Act, information must be provided in relation to the injuries and diseases that occur as a result of work.

Employee Duties

An employee must inform the employer, either orally or in writing, of any injury caused by an accident. This should happen as soon as practicable after the accident. A completed entry in an accident book is sufficient notice of the accident.

Investigation

The employer must investigate the circumstances of every accident that is reported and record any discrepancy from the employee's report.

Record-keeping

Every employer must keep an accident book readily accessible in which the appropriate particulars of any accident causing injury to an employee may be entered. All accident books have to be kept for three years

The prescribed form of record required under these regulations is an Accident Book which contains the following information:

(a) full name, address and occupation of injured person;
(b) date and time of accident;
(c) place where accident happened;
(d) cause and nature of injury; and
(e) name, address and occupation of person entering the details, if other than the injured person.

In 2003 a new accident book was introduced to comply with data protection legislation. It is designed so that any individual recording an accident is unable to access personal details of previous records. There is no legal requirement to use the prescribed accident book as long as the appropriate records are kept and retained. Where an employer chooses not to use the prescribed accident book, they must, of course, ensure that they keep personal details confidential.

REPORTING OF INJURIES, DISEASES AND DANGEROUS OCCURRENCES REGULATIONS. (OFTEN KNOWN AS RIDDOR)

For serious incidents at work the employer is required to notify either the local authority or the Health and Safety Executive (HSE).

Incidents requiring immediate reporting
Major injury or fatality

We must report quickly, usually by telephone supported by a form later, any fatality or major injury arising from or in connection with work. This includes accidents to a person who does not work on the premises (customer, patient, visitor etc.).

Major injuries include:

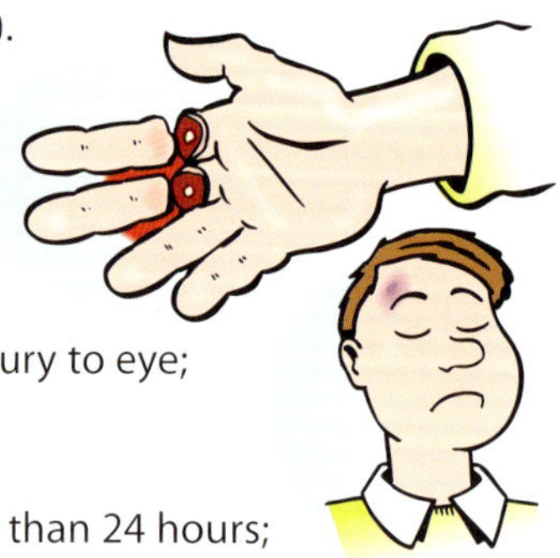

- fracture other than fingers, thumbs or toes;
- any amputation;
- dislocation of shoulder, hip, knee or spine;
- loss of sight (temporary or permanent);
- chemical or hot metal burn to eye or penetrating injury to eye;
- unconsciousness due to electric shock;
- any injury requiring resuscitation;
- any injury requiring admittance to hospital for more than 24 hours;
- loss of consciousness due to asphyxia;
- acute illness or loss of consciousness due to inhalation of any chemical; and
- acute illness caused by exposure to infected material.

Dangerous occurrences

We must also report without delay a number of serious incidents that do not cause injury. The full list is in the regulations, but includes:

- collapse, overturning or failure of load-bearing parts of lifts and lifting equipment;
- explosion, collapse or bursting of any closed vessel or associated pipework;
- failure of any freight container in any of its load-bearing parts;
- plant or equipment coming into contact with overhead power lines;
- electrical short circuit or overload causing fire or explosion;
- any unintentional explosion, misfire, failure of demolition to cause the intended collapse, projection of material beyond a site boundary, injury caused by an explosion;
- accidental release of a biological agent likely to cause severe human illness;
- collapse or partial collapse of a scaffold over five metres high, or erected near water where there could be a risk of drowning after a fall;
- unintended collapse of any building or structure under construction;
- alteration or demolition where over five tonnes of material falls; a wall or floor in a place of work; any false-work;
- explosion or fire causing suspension of normal work for over 24 hours; and
- accidental release of any substance which may damage health.

Incidents that require reporting within 10 days

Over three-day injuries

Employers must notify within 10 days work-related ill health or injury that causes an employee to be off work or not able to perform their usual job for more than three days. These are reported on Form F2508.

Occupational disease

There are also many occupational diseases that must be reported on form F2508A.

There are three main categories of disease:

A) Those diseases associated with the physical agents or the physical nature of the work, including:
- disease associated with exposure to ionising radiations; and
- cramp of hand or arm due to repetitive movements, bursitis or inflammation of arm or hand tendons when associated with work.

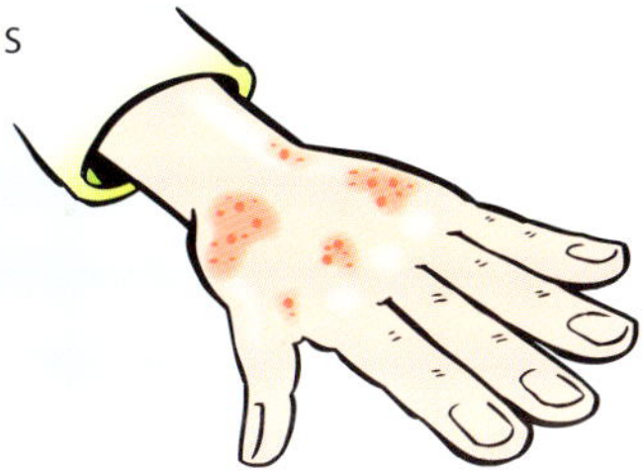

B) Infections due to biological agents including:
- hepatitis when associated with work;
- TB when associated with work; and
- any infection reliably attributable to work.

C) Conditions due to substances including:
- poisoning due to metals, oxides of nitrogen etc.;
- occupational dermatitis due to exposure to agents at work; and
- occupational asthma due to exposure to agents at work.

Reporting

Reporting can be:

- by telephone 0845 300 99 23
- by fax 0845 300 99 24
- over the internet at www.riddor.gov
- by post to Incident Contact Centre, Caerphilly Business Park, Caerphilly, CF83 3GG
- by email to riddor@natbrit.com

Employers must investigate the circumstances, and keep records, of all reported incidents.

EMPLOYER'S POLICY AND PROCEDURE

In practice the employer needs to have a robust system to ensure that the correct information is available at the right time.

This should consist of detailed procedures in the company health and safety manual. The procedure should include clear provisions for the accident book and local incident form to be completed, where they are located and what happens to the forms. Who completes the RIDDOR forms to notify the incident and how they are handled must be clearly defined.

The procedure needs to be part of the general health and safety training provided for all employees.

USE OF INFORMATION

Many employers use a computerised information handling system for incident reports. This enables them to track and identify trend for incidents for the organisation and locally, allowing risk assessments and training plans to be amended accordingly.

Over a period of time the information will be a useful source of data for internal comparison of performance - but beware! If targets are set for improving performance by reducing numbers of accidents reported, there is a temptation **not** to report. It is not usual to directly compare different organisations on rates of reported incidents as this only compares the accuracy and use of the reporting system and not the accident rate.

It is also necessary to have complete control of information on incidents and their investigation, as litigation is becoming common. The ability to provide timely information for legal advisors and insurers is a very important part of claims management, and can itself reduce premiums considerably as it allows the early settlement or challenging of claims.

ACCIDENT STATISTICS

The Health and Safety Executive regularly publish statistics on the reporting of fatalities, major injuries and occupational disease. All the information is on their website www.hse.gov.uk/statistics.

Over the last ten years there has been a trend of decreasing numbers of deaths at work, with 226 reported during 2002/03. There were 28,426 major injuries and 126,004 three-day injuries reported during this period.

Of the major injuries, 37% were due to slips, trips and falls and 12% due to manual handling.

Of the three-day injuries, 39% were manual handling and 24% slips, trips and falls.

However we must remember that studies have shown that the HSE only receive reports of 40% of incidents that should be reported.

CAUSES OF ACCIDENTS

There are three sets of factors that contribute to safety:

1. human factors – behaviour and ability, what we do and why we do it;
2. occupational factors – the job or task we do; and
3. environmental factors – the surroundings in which we work.

Sometimes accidents happen because these factors occur at the same time, and they all contribute to the outcome.

1. HUMAN FACTORS

There has been a great deal of research into why we make mistakes, including the following:

- Lack of knowledge:
  everyone performing a job should be given information, instruction and training to do the job safely. This should be easily understood.
- Lack of concentration:
  tiredness, monotony, repetitive tasks, distraction, alcohol and drugs may all contribute to the worker losing concentration. Long working hours are an example, as is trying to work after drinking alcohol.
- Horseplay:
  the workplace is usually not a suitable area to fool around in.
- Lack of capability:
  sometimes we train people, and they still cannot do the job safely. Sometimes due to their health and physical fitness, their ability is limited.

2. OCCUPATIONAL FACTORS

These are directly associated with the job or task, including:

- manual handling of loads, particularly if they involve repetitive tasks and bending, stretching or twisting;
- the use of equipment such as ladders, hand tools, computers etc.; and
- using hazardous substances.

3. ENVIRONMENTAL FACTORS

These are associated with the area in which we work and include:

- safe entry and exit;
- safe floors;
- adequate space to do the job safely;
- noise;
- heating, lighting and ventilation; and
- somewhere to change, eat and rest.

An example of poor manual handling, whilst using a ladder.

3 Principles of Health & Safety Law

Health and safety law can be complex and difficult to understand. We have rights as individuals that have developed over many hundreds of years within common and civil law. Since the Industrial Revolution, a large number of acts and regulations have been put in place to protect workers, making health and safety a criminal matter.

COMMON AND CIVIL LAW

Every employer is under a common law obligation (irrespective of legislation) to take all reasonable care to prevent injury to employees or damage to their health. This is usually carried out by:

a) selecting employees competent for the task;
b) providing proper plant and equipment; and
c) instituting and maintaining a proper system of work and effective supervision.

The remedy for injury or loss by breach of this obligation is the award of damages. The plaintiff can be an employee or other person.

Civil law is also concerned with the award of damages to a complainant. To be successful, the complainant must prove three things:

a) that the defendant owed him a duty of care;
b) that the defendant broke that duty, i.e. was negligent; and
c) that the defendant's negligence caused the complainant damage.

A common way of proving negligence is to show that the defendant did not comply with a statutory requirement. It is important to note that the burden of proof in civil cases is 'On the balance of probability', not 'beyond reasonable doubt' as in criminal cases.

Employer's liability insurance

The Employer's Liability (Compulsory Insurance) Act 1969 requires all employers to carry insurance that will allow an employee to claim even if the employer has gone out of business. The cover has to be for a minimum of £5 million and certificates must be displayed and retained for 40 years.

Woolf reforms

Until the late 1990s many claims took years to be settled, and legal costs became prohibitive, making it very difficult for the majority of workers to pursue a claim for fear of failing.

Lord Chief Justice Woolf introduced some reforms to try and ensure that claims are now settled within a set period of time and that costs are kept to a minimum.

The Pre-Action Protocol

To speed up civil claims, Woolf introduced the Pre-Action Protocol, a prescribed process, with time limits. It begins with a standard letter of claim from the claimant. This letter should contain enough information, with a description of the accident and details of why they believe the employer has been negligent, to substantiate a realistic claim. There also has to be a broad valuation of the claim.

The letter of claim must be replied to within 21 days, identifying the insurer (if any). If the employer does not reply within 21 days he is likely to be liable for all the costs, both to himself and the claimant.

The employer has a maximum of three months to investigate. By the end of three months, there has to be a response stating whether or not liability is denied and, if so, giving reasons plus documentary evidence of why liability is denied. There is a list of documents that are likely to be required as evidence, including statements, risk assessments, training records etc.

Where liability is admitted, the presumption is that the defendant is bound by this admission for all claims up to £15,000.

Managing claims

Employers find it a benefit to have a claims management procedure, detailing action that must be taken as soon as the letter of claim is received. Employers and insurers need to investigate quickly and to arrive at the correct decision on liability. The majority of the fast track claims will be dealt with, including litigation, within a period of six months.

The employer and insurer should be able to decide quickly on whether to settle or not, and, if not, provide evidence on why liability is denied.

CONTRIBUTORY NEGLIGENCE - evidence sometimes shows that the complainant was partially the cause of the loss or damage. The court may rule in these cases that the compensation sum be reduced by the proportion of the blame attached to the complainant.

Civil Courts

Although cases are often agreed before reaching court, common law cases are taken to County Court and occasionally the High Court.

CRIMINAL LAW

The sources of criminal law are case law and statute law.

Case Law

Case law originated from the need to make judgements consistent. A hierarchy of courts was built up with a reporting system of cases so that lower courts follow the decision of higher courts. The more simple cases are dealt with at Magistrates' Courts. If a jury is needed, the trial will be at the Crown Court. In Scottish Law the Procurator Fiscal decides whether a prosecution is in the public interest and criminal prosecutions take place in the Sheriff's Court.

Origins of Statute Law

Statute law originates in three ways:

a) Acts of Parliament - e.g. The Health and Safety at Work etc. Act 1974

The process of making an Act of Parliament is very complex. A bill may originate in either house (House of Commons or House of Lords), usually introduced by the minister responsible, shortly after the Queen's Speech. This is known as the first reading.

At the second reading the general principles are debated and voted on. If accepted, it goes to committee stage with usually a small committee of MPs. Each clause is now debated.

The report stage debates the committee report and then whole bill is debated at the third reading. The bill is then passed to the other house where it must follow the same procedure. If passed by both houses the bill is given Royal Assent and becomes an Act.

This process often takes almost a full year.

b) Regulations – e.g. The Management of Health and Safety at Work Regulations 1999

An Act will often include a section allowing a minister to make regulations on a specific topic. This allows very detailed law to be made and if necessary, amended quickly. Another example is the Health & Safety (Display Screen Equipment) Regulations.

c) EU

In the field of health and safety, EU legislation is the usual starting point for UK law.

The European Commission send proposals to the European Council for Directives to be made. If a majority of the member states agree, the directive is made and each country is required to put the provisions of the directive in domestic legislation within a time limit. The 'six pack' of regulations made in 1992 are examples of legislation originating in Europe.

Approved Codes of Practice

Regulations are usually published with advice and guidance on how to implement them within the workplace. Sometimes this is known as an 'Approved Code of Practice' (ACOP). This means that the advice within it has been accepted by the Health and Safety Commission. If the employer does **not** do what is advised in the ACOP, if prosecuted the employer would have to prove that he has done something equally as effective or he will be found guilty of an offence. The ACOP is not law, but it is accepted as being very near to it.

The Health and Safety at Work etc. Act 1974

This Act changed the legal approach to health and safety by introducing for the first time a comprehensive system of placing duties on employers, employees, designers and manufacturers. This made the law self-regulating, replacing an enforcement-led system of specific, and often industry-based, regulations.

The major parts of Act are:

Section 2. This section requires employers to ensure, as far as is reasonably practicable, the health, safety and welfare of all employees. In particular this includes the provision of:

- safe plant and safe systems of work;
- safe handling, storage and transport of articles and substances;
- necessary information, instruction, training and supervision;
- a safe place of work, with safe access and egress; and
- a safe working environment with adequate welfare facilities.

Section 2 also includes the requirements for a health and safety policy and for consultation with employees.

Section 3 requires employers to ensure as far as is reasonably practicable, the health and safety of others. This includes visitors, customers, contractors etc.

Section 7 requires employees to take reasonable care of themselves and others and to co-operate with their employer.

Section 8 makes an offence of recklessly or intentionally interfering with or misusing anything provided in the interests of health and safety.

Sections 37 & 38 provide for the personal prosecution of managers in certain circumstances. This can be as well as or instead of the employer, if the offence was due to their consent, connivance or neglect.

'As far as reasonably practicable'

This is included in The Health And Safety At Work Act to introduce achievable control. Case law has indicated that this includes cost. It also allows change over time. In other words, the level of risk is equated with time, effort and costs. The employer does not have to do *everything* possible to prevent risk.

'Reasonable' has been interpreted as the 'act of a reasonable man, what the man on the Clapham omnibus would do'. This indicates that what is 'reasonable' may change over time, as society and technology change.

Enforcement

There are three types of health and safety inspectors:

- Health and safety inspectors are employed by the government body, the Health and Safety Executive (HSE), and are responsible for factories, schools, hospitals, agriculture etc.;
- the local authority, usually environmental health staff, is responsible for shops, offices, warehouses etc.; and
- fire authorities are responsible for fire safety in workplaces.

To allow the inspectors to carry out their work effectively, they are given many powers in The Health and Safety at Work Act. Inspectors can:

- gain access to any workplace at any time;
- take equipment or materials into premises to assist investigations;
- carry out any investigations;
- direct that locations remain undisturbed;
- take measurements, photographs and samples;
- order removal and testing of equipment;
- take away articles or equipment for testing or examination; and
- take statements, records and documents etc.

When carrying out an inspection, the inspector will ask questions. They will also look at documents such as policies and procedures and risk assessments and observe activity in the workplace. On finding a contravention of legislation, inspectors have five options:

- give advice, usually in writing – this is by far the most common action;
- serve an improvement notice on the owner, employer etc.;
- serve a prohibition notice on the owner, employer etc.;
- commence criminal proceedings against the owner, directors, managers, etc; and
- seize, render harmless or destroy any article which they consider to be the cause of imminent danger.

Inspectors will always confirm action taken with employees' representatives.

Improvement Notices

These are served by the inspector, usually on the employer, when a breach of a statutory requirement has occurred. The notice gives time limits for correcting the omission, and action that needs to be taken to comply with the relevant legislation. These are often served when advice has not been taken. It is an offence not to comply with an improvement notice within the time limit. An appeal may be made to an employment tribunal within 21 days, and the notice is suspended during the appeal.

Prohibition Notices

This notice is also served by the inspector, in this case when there is a serious risk of personal injury. It may require work to cease immediately or may be deferred. The duration of the notice and action necessary are detailed on the notice and it remains in force until the inspector removes it. There is an appeal to a tribunal within 21 days, but this notice remains in force until the appeal. It is a serious offence not to comply with a prohibition notice.

All health and safety notices are available for the public to see on the HSE website *www.hse.gov.uk.*

Penalties

Following successful prosecution the penalties can be severe:

- **Magistrates' Court**
  Up to £5,000 per offence against a regulation.
  Up to £20,000 per offence against The Health and Safety at Work Act or breach of an improvement or prohibition notice.
  Up to six months in prison.

- **Crown Court**
  Unlimited fine.
  Up to two years in prison.

It has been known for a defendant to plead guilty in the Magistrates' Court and because of the seriousness of the offence be referred to the Crown Court for sentence.

4 Managing Health & Safety at Work

It is now clearly recognised that if health and safety is managed as any other part of a business is managed, in a systematic and planned manner, success in reducing the costs associated with poor control will be achieved. It is therefore no surprise that there is legislation in place requiring all employers to manage health and safety in an effective and systematic manner.

THE MANAGEMENT OF HEALTH AND SAFETY AT WORK REGULATIONS 1999

These regulations embrace all aspects of management, including information, instruction, training, competence, contractors etc. They were published with an accompanying Approved Code of Practice (ACOP) which gives practical guidance on the provisions. Failure to comply with this code is not an offence, although it is clear that failure is likely to be taken by a court as proof that there has been contravention of the regulations. It will then be necessary to prove that compliance with the law has been achieved in some other way.

(The regulations include the general requirements of risk assessment; these are covered in Chapter 5 of this book.)

Many health and safety professionals regard regulation 5 to be the single most important piece of health and safety law. It requires employers to ensure that health and safety is part of the management 'culture' of the organisation, that it is part of the way that work is done, not an extra part of the job or task.

Regulation 5 Every employer shall make and give effect to arrangements for:

- planning;
- organisation;
- control;
- monitoring; and
- review of protective and preventive measures.

ACOP - These arrangements should be integrated with the management system for other purposes.

The arrangements are effective management systems as described in the HSE publication 'Successful Health and Safety Management'. ISBN 0.71761276.7 (£12.50) from HSE Books.

Planning - the employer should adopt a systematic approach to the completion of risk assessment, decide on priorities and set objectives. A programme, with deadlines, for the completion of assessments and the implementation of measures should be devised.

Organisation - involve employees and representatives in risk assessment, establish effective communication, provide adequate information and make sure that employees are competent to do their job safely.

Control - clarify responsibilities, ensure co-ordination and ensure everyone understands their responsibilities. Give them time to do it and set standards of performance that are reviewed. Ensure adequate supervision.

Monitoring - measure performance with effective supervision and a systematic and planned programme of inspections. Report and investigate incidents.

Review - establish priorities for action following monitoring, and review the whole management system.

HEALTH AND SAFETY POLICIES

Legal requirements

Section 2(3) of the HSWA states: 'It shall be the duty of every employer to prepare a written statement of his **general** policy with respects to the health and safety at work of his employees and the organisation and arrangements for the time being in force for carrying out that policy, and to bring that statement and any revision to the notice of all his employees.'

The essential elements of the policy are therefore:

a) policy to demonstrate **commitment;**
b) **organisation** to clearly state people, their duties and accountability; and
c) **arrangements** - systems and procedures that ensure that work is carried out safely.

Commitment

The writing and updating of a safety policy should give the employer the vehicle to maintain the impetus of health and safety at work. It should demonstrate that the business accepts that health and safety is an integral part of its organisation at all levels. The commitment from the highest level of management is demonstrated by the signing of the policy by the chief executive, board of governors etc.

Each policy is unique and is a formal declaration that the company intends to devote sufficient effort and resources to the organisation of health and safety. All employees, visitors and contractors should easily understand the policy.

Organisation

Although ultimate responsibility lies at the top, duties have to be delegated and this forms the link between policy and organisation. There should be clear and unambiguous 'shape' to health and safety management, usually done diagrammatically. Everyone should understand their role and responsibilities for health and safety at work and everyone should know who is responsible for their health and safety. It is also helpful for this part of the policy to detail how individuals are accountable.

Indicators of a successful 'organisation' are:

A) unbroken and logical delegation of duties to supervisor level. All employees should be able to recognise where responsibility for their job lies;
B) definitions of roles, clarified in individuals' job descriptions;
C) identified and adequate support - safety professional, occupational health service etc.; and
D) details of supervisors who have appropriate skills and knowledge within their area of work.

Arrangements

These should represent an honest appraisal of how people work and not how managers think people should work. Included in successful policies are, for instance, arrangements for contractors on site, visitors to the site, out of hours working etc.

Typical arrangements necessary for a successful policy are:

A) planning prior to any change in circumstances, contract placement, etc.;
B) a systematic process of documented risk assessment with prioritisation of action;
C) specific instructions for particular hazards (safe systems of work);
D) training policy and plan implemented, recorded and monitored;
E) occupational health policy and service provided;
F) emergency and first aid procedures documented; and
G) safety manual documented and maintained.

THE ROLE OF THE LINE MANAGER AND SUPERVISOR

The health and safety policy and manual include all the necessary detail about how jobs are completed safely, including how staff are to be trained. But we know that policies are not always read and understood, and that individuals make mistakes, take short cuts etc. It is the line manager and/or supervisor who ensures that these safe systems of work are followed in the workplace.

It is difficult to define who is a line manager or supervisor, as different organisations often use different job titles. However, it is generally accepted that if you are directly responsible for a member of staff, you supervise them. For a new member of staff, recently appointed and with limited experience, this may include more direct and personal supervision than for someone who has been employed for some time and has been seen to work in a safe way.

The supervisor ensures that the policy, procedures and safe systems of work are consistently followed by everyone. There are requirements in The Health and Safety at Work Act and many regulations for effective supervision of employees.

The role of the supervisor includes:

- risk assessment;
- monitoring performance;
- training;
- instruction;
- emergency procedures; and
- consultation and involvement.

We all recognise that many supervisors also have their own job to do, making these roles more difficult to complete.

HEALTH SURVEILLANCE

This is the monitoring of the health of an individual to ensure that work is not affecting their health. It should be carried out at an early stage, preventing further harm.

The need is identified by risk assessment, using information about the job that often originates from HSE advice or guidance.

Health surveillance confirms that safe systems of work are effective and identifies individuals who are at increased risk because of the nature of their work.

Examples of health surveillance include:
- eye and eyesight testing for display screen users;
- checking of hands for dermatitis;
- checking levels of lead in blood for lead workers;
- lung and breathing checks for users of substances that may cause occupational asthma; and
- hearing checks for those who work in noisy environments.

Surveillance is often carried out by occupational health staff, and it must result in a health record.

Regulation 6 **Every employer shall ensure that employees are provided with appropriate health surveillance.**

ACOP - The risk assessment carried out under Regulation 3 (amongst others) will identify risks to health and safety which require health surveillance, either by legislative requirement (e.g. COSHH) or because the following criteria apply:
a) *there is an identifiable disease or health condition related to the work concerned;*
b) *valid techniques are available to detect indications of the disease or condition;*
c) *there is a reasonable likelihood of the disease or condition occurring; and*
d) *surveillance is likely to help protect the health of the employee.*

COMPETENCE

This regulation requires all employers to have access to advice on health and safety matters. The definition of competence is important, and includes:
- knowledge – usually qualifications in health and safety at work;
- skill – to implement the knowledge effectively; and
- experience – in the employer's type of workplace.

Regulation 7 **Every employer shall appoint one or more competent persons to assist in taking steps to comply with statutory requirements.**

ACOP - Employers are responsible for ensuring that the person appointed for this function is competent, and that they are given adequate information and support.

EMERGENCY SITUATIONS

These regulations apply to fire safety and other situations such as working in confined spaces, underground, with hazardous substances etc.

Most organisations have emergency and contingency planning in place as part of systematic risk management. This will include how to contact emergency services, call-out procedures for key staff, training etc.

Regulation 8 **Every employer shall establish procedures to be followed in the event of serious or imminent danger to persons at work.**

This includes the nomination of competent persons to assist evacuation.

Employers must ensure that there is restricted access to areas that have significant risk.

Regulation 9 Employers shall ensure that necessary contacts with external services such as first aid, medical, fire, police are made.

Procedures should be in place to contact services when necessary and clarify when work may resume after the emergency.

INFORMATION FOR EMPLOYEES

With employees from many diverse backgrounds and abilities the employer must ensure that information provided for staff is understandable and understood.

Part of the recruitment process should include knowledge of their literacy, and the steps to be taken to provide information that is easily understood by the new employee. Literacy is not just about the ability to read and write English, it includes the need to understand information provided. An example is that many policies and procedures are produced in a style of English that is difficult to follow. We should all try to write policies that are simple and easy to understand.

Regulation 10 Every employer shall provide his employees with comprehensible and relevant information on risks to his health and safety and the preventive and protective measures.

ACOP - Information must be provided in a form that is easily understood by the recipient. Care must be taken with employees that have difficulties with reading and when English is not the first language.

Information must also be provided for the parents of employed children (those below school leaving age).

EMPLOYERS THAT SHARE A WORKPLACE

Sometimes a workplace is shared, e.g. an office block used by many businesses. In this case there should be co-operation in matters such as cleaning, fire safety etc.

Regulation 11 Where two or more employers share a workplace, each employer shall co-operate with other employers as necessary to comply with statutory requirements.

Each employer shall take all reasonable steps to co-ordinate measures taken by him with measures taken by others.

CONTRACTOR SAFETY

There have been many examples of prosecutions of employers who have employed contractors who have had an accident. The basic requirement is for the client to have some control of the contractor's activities.

Many employers have a policy and procedure for the management of contractors. This will include matters such as the contractor's insurance cover and their safe systems of work. It is usual for contractors to provide clients with a 'method statement'. This details how and when the contractor will carry out the job. The client is then able to inform his employees of the risks and control measures. The client must carry out reasonable checks to make sure that the contractor is abiding by the measures in the method statement.

Typical contractors include: decorators, window cleaners, caterers, cleaners, engineers, builders etc.

Large building construction contracts are covered by specific legislation and The Construction (Design and Management) Regulations 1994.

Regulation 12 **Every employer shall ensure that the employer of any employees from an outside undertaking who are working on his premises is provided with comprehensible information on the risks of working in that workplace and preventive and protective measures taken.**

Risk assessments carried out under Regulation 3 will have identified risks to employees covered by this regulation. The host employer must take steps to ensure that the second employer's staff receive information on risks and protective and preventive measures. Adequate checks must be made to ensure that information is passed on.

The second employer should also inform the first employer of any risks introduced during the work carried out.

CAPABILITY AND TRAINING

Staff training is the subject of Chapter 12, but the major legal requirement lies in The Health and Safety at Work Act and the Management Regulations.

Firstly, the employer must not give an employee a job or task until they are satisfied that they are competent and capable to do that job safely. This is significant in that the employer should therefore test and record competence, making training records vital in defending claims.

Regulation 13 **Every employer shall take into account their employees' capabilities when entrusting tasks to employees.**

Every employer shall ensure that employees are provided with adequate health and safety training:

a) on recruitment;
b) on being exposed to increased risk because of change of job or the use of new equipment.

Training shall be repeated periodically where appropriate and take place in working hours.

ACOP - Employers should not allocate tasks that are not within the capability of the employee.

Training needs are greatest on recruitment and should cover first aid, fire, and evacuation procedures. Changes in the working environment may require further training. For larger employers, effective management of training requires a training plan and systematic review. Employees are not required to pay for their own training (see section 9 of the Health and Safety at Work Act (HSWA). Training needs are usually identified in risk assessments.

EMPLOYEES' RESPONSIBILITIES

As in the Health and Safety at Work Act, employees have legal responsibilities under the Management of Health and Safety at Work Regulations (MHSWR) to 'do as they are told'.

The reporting of faults is important, and the employer should have a procedure for this to occur and action to be taken to correct the problem.

Regulation 14 **Every employee shall use equipment or safety devices in accordance with training provided by the employer.**

Every employee shall inform their employer of any serious or imminent danger to health and safety of any other employee.

ACOP - This reinforces the provisions of section 7 of the HSWA, although it does not reduce the employer's responsibility to comply with duties under these and other regulations.

5 Risk Assessment

Risk assessment was introduced into health and safety legislation with the original Control of Substance Hazardous to Health Regulations. Since then it has become commonplace in legislation, in particular with the introduction of the 'six-pack' of European based regulations enacted in January 1993.

The purpose of risk assessment is to help the employer determine what measures should be taken to comply with the employer's duties under The Health and Safety at Work Act 1974.

The principles of risk assessment are the subject of the Approved Code of Practice (ACOP) issued with The Management of Health and Safety at Work Regulations 1999. The regulations require that employers undertake a **systematic general** examination of their work activity and that they should record the **significant findings** of that risk assessment.

It is important to remember that a risk assessment is only a piece of paper and it is how the control measures are implemented, monitored and reviewed that makes a job safe. The line manager or supervisor's role in this is vital.

PRINCIPLES OF RISK ASSESSMENT

A risk assessment consists of five steps.

1. IDENTIFY HAZARDS
2. IDENTIFY THOSE WHO MAY BE HARMED
3. EVALUATE RISK BY CONSIDERING EXISTING PRECAUTIONS
4. RECORD THE FINDINGS
5. REVIEW AND REVISE AS NECESSARY

A hazard is something with the potential to cause harm; risk is the likelihood of harm.

The ACOP supporting The Management of Health and Safety at Work Regulations 1999 says that trivial risks can be ignored, unless the work increases those risks or there is significant relevance to the particular work activity. To be in a position to concentrate on significant risks, the hazards must be identified for all work activities. To aid this process employers often use checklists that will fulfil the need for employers to record the findings of the risk assessments.

RISK ASSESSMENT FLOW CHART

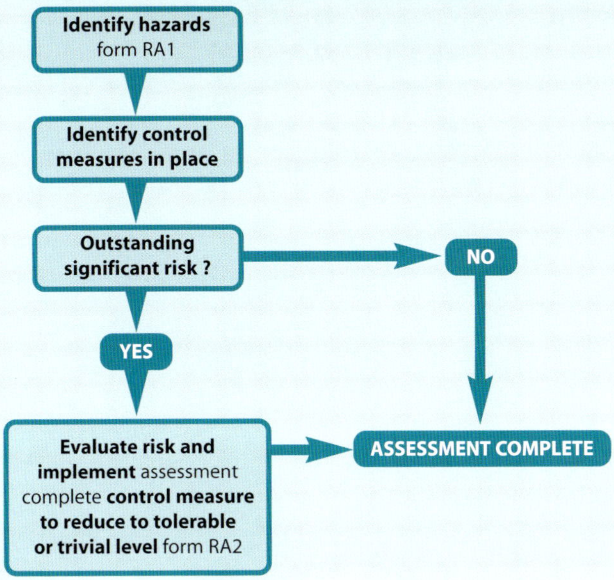

The flowchart explains the questions that need to be answered in a risk assessment.

RISK ASSESSMENT - IDENTIFYING HAZARDS, EVALUATING AND PRIORITISING RISK

This is likely to involve the use of a checklist, and an example is given as RA1. This is used to identify hazards for the activity being assessed. The steps currently taken to prevent the hazard being a risk and any measures that can be introduced simply and quickly are also recorded on this form. It is important to identify those who may be harmed.

If there is no significant risk as the current control measures are judged to be adequate, the risk assessment is finished.

If there is a significant risk, we must evaluate it and decide what measures need to be taken to reduce the risk to an acceptable level. There are many ways of evaluating risk, but it is generally accepted that a matrix be used. The most simple is probably that detailed in BS 8800. A form such as the example given as RA2 can be used; it is a checklist that helps evaluate the risk and prioritise action.

	SLIGHTLY HARMFUL	HARMFUL	EXTREMELY HARMFUL
HIGHLY UNLIKELY	Trivial risk	Tolerable risk	Moderate Risk
UNLIKELY (could occur sometime)	Tolerable risk	Moderate risk	Substantial risk
LIKELY (not surprised if it did happen)	Moderate risk	Substantial risk	Intolerable risk

An important part of RA2 is that the evaluation of risk is carried out using this matrix. The level of risk before and after the implementation of control measures must be clearly marked. The decision on the level of risk is usually taken after consultation with those staff at risk - they will have a clear idea of the risks and what can be done to reduce risk to an acceptable level.

Guidance on evaluating risk

All evaluation of risk is a factor of likelihood of harm and consequence.

To determine likelihood, we have to consider the adequacy of control measures currently implemented. In addition, it may be significant to take into account the number of people exposed to the risk, and their vulnerability. An incident may be more likely if more staff are at risk and it may be necessary to carry out a separate assessment of risk for staff who may be exposed to additional risk due to their personal knowledge, skills or capacity.

Level of harm

Consequence can be classified by the degree of harm. **Slightly harmful** can be taken as superficial injuries requiring a first aid treatment, a nuisance, irritation or temporary discomfort; **harmful** as lacerations, burns, concussion, sprains, minor fractures, dermatitis etc. and resulting in lost time from work: **extremely harmful** as resulting in death, amputations, major fractures, significant life shortening illness or major injury to one or more people.

Likelihood

When taking decisions on likelihood, do not forget any vulnerable groups of staff, such as lone or peripatetic workers and those who work at night or are 'on-call'. These groups of staff or individuals may need a separate assessment.

Highly unlikely can be taken as probability close to zero, **unlikely** as an event that could occur sometime, and **likely** as an event that you would not be surprised if it did happen.

ACTION RESULTING FROM THE EVALUATION OF RISK

Control measures are implemented to reduce the risk to a tolerable or trivial level. The action required by the evaluation of the risk is detailed below. The evaluation of risk in this manner helps a business plan to prioritise action. This is particularly helpful when comparing risks from different situations. For example, is the risk from using a particular machine a higher priority than personal safety?

RISK	ACTION AND TIMESCALE
Trivial	No action is required and no documentary records need to be kept
Tolerable	No additional controls are required. Consideration may be given to a more cost-effective solution or improvement that imposes no additional cost burden. Monitoring is required to ensure that controls are maintained.
Moderate	Efforts should be made to reduce the risk, but costs should be carefully measured and limited. Risk reduction measures should be implemented within a defined time limit. Where a moderate risk is associated with extremely harmful consequences, further assessment may be necessary to establish more precisely the likelihood of harm as a basis for determining the need for improved control measures.
Substantial	Work should not be started until the risk has been reduced. Considerable resources may have to be allocated to reduce the risk. Where the risk involves work in progress, urgent action should be taken.
Intolerable	Work should not be started or continued until the risk has been reduced. If it is not possible to reduce risk even with unlimited resources, work has to remain prohibited.

The matrix, definitions and suggested actions are taken from BS8800 'Occupational Health Systems'.

The date of the implementation of measures is recorded when they have all been fully implemented. It is acknowledged that there may be some time before all measures can be dealt with. In this case, outstanding risks can be entered on a risk register and the highest risk items actioned as a priority. If this is so there must be a record of why the measures have not been implemented and steps that are being taken to remove or reduce risk in the meantime.

Communication with employees

It is a requirement of the regulations that staff be trained on recruitment, on change of responsibilities, the introduction of new equipment or new system of work. The training should be repeated periodically, adapted to the capabilities of the worker concerned and take place during working hours. It is also a requirement that the employer provides all employees with comprehensive and relevant information on the details of risk assessment and the measures to be implemented.

Staff involved in the activity must be consulted during the assessment of risk, and their training records must include the objectives that were met during the training, that they understood the training and how competence was tested. (This is usually a record that the employee has been observed by the trainer to complete the task in a satisfactory manner.)

USE OF RISK ASSESSMENTS

The regulations require risk assessments to be written if there are more than 5 employees. They have to be available for all employees to see, not locked away in a cupboard.

Everyone understands that risk assessments are not everyday reading - but they should explain what is done to make the job safe. It is how the assessment is translated into action that does matter, the training, instruction and supervision that implement the control measures.

When there are a number of outstanding risk assessments, employers usually put together a 'Risk Register'. This is a simple list with the highest risk issues at the top, and allows everyone to understand how prioritisation takes place.

RA1

GENERAL RISK ASSESSMENT *THE FUNCTION OF THIS DOCUMENT IS TO IDENTIFY HAZARDS IN THE WORKPLACE AND TO DETERMINE WHICH HAZARDS PRESENT A RISK*

DEPARTMENT	LOCATION	DATE (this assessment)	DATE (last assessment)

Hazard	existing measures	is hazard a risk ? YES or NO	new measures required or go to RA2	persons at risk

assessors signature name of competent person other assessors name (print)

RA2 RISK ASSESSMENT

THE FUNCTION OF THIS DOCUMENT IS TO EVALUATE THE RISKS TO HEALTH AND SAFETY OF A HAZARD, TO IDENTIFY CONTROL MEASURES AND RECORD IMPLEMENTATION

DEPARTMENT	LOCATION	DATE	DATE MEASURES IMPLEMENTED

FULL DESCRIPTION OF HAZARD	**EVALUATION OF RISK**		**RING ONE BOX ONLY**		CONTROL MEASURES TO BE TAKEN
		SLIGHTLY HARMFUL (first aid)	HARMFUL (time off work)	EXTREMELY HARMFUL (major injury or death)	
	HIGHLY UNLIKELY	trivial risk	Tolerable risk	moderate risk	
	UNLIKELY (could occur sometime)	tolerable risk	Moderate risk	substantial risk	
	LIKELY (not surprised if it did happen)	moderate risk	Substantial risk	intolerable risk	
ASSESSOR SIGNATURE AND PRINT	***tolerable risk - monitor compliance***	***moderate risk - plan control measures***	***Substantial risk - take urgent steps***	***intolerable risk - stop***	level of risk after implementation of measures
					SIGNATURE FOR EMPLOYER

LEGAL REQUIREMENTS

The Management of Health and Safety at Work Regulations 1999 and Approved Code of Practice detail how risk assessments should be carried out.

Risk Assessment

Regulation 3 **Every employer must make a suitable and sufficient written assessment of the risks to health and safety of his employees to which they are exposed at work.**

Every employer must make suitable and sufficient written assessments of the risks to the health and safety of others arising out of or in connection with the conduct by him of the undertaking.

The assessment shall be reviewed if it is no longer valid, or if there has been significant change.

ACOP - The assessment shall be recorded if there are more than 5 employees.

A suitable and sufficient risk assessment should:

- **a)** *identify the hazards (potential to cause harm);*
- **b)** *evaluate risks (by considering likelihood of harm, severity of harm and the number of people exposed) arising out of work. Trivial risks can be ignored;*
- **c)** *determine measures necessary to comply with the HSWA;*
- **d)** *be appropriate to the nature of the work and valid for a reasonable period of time;*

The assessments should be reviewed and if necessary modified on a regular basis or after a near miss or accident.

Hierarchy of control

There is a requirement to follow a systematic approach to deciding which control measures to implement. The most important is to try and avoid the risk altogether. If this is not possible, we evaluate the risk and put in place a safe system of work that takes account of the individual, technological progress etc.

Personal protective equipment (PPE) should only be used if there is no other way of protecting employees against the risk (see Chapter 15).

Regulation 4 **Where an employer implements preventive or protective measures, they shall follow the principles in Schedule 1 of the regs.**

Following the assessment of risk, the employer does have to take preventive and protective measures. The following principles need to be applied:

- **a)** *if possible avoid the risk all together;*
- **b)** *evaluate risk;*
- **c)** *combat risks at source;*
- **d)** *adapt work to the individual wherever possible;*
- **e)** *take advantage of technical progress to make work safer;*
- **f)** *replace dangerous with the non- or less dangerous;*

g) *adopt sound management principles;*
h) *give priority to those measures that affect the whole workplace; and*
i) *ensure workers understand what they need to do.*

Young people

There are requirements to ensure that young people (between the ages of 16 and 18 years) work safely. This includes the carrying out of risk assessments that take into account their experience of work and maturity.

Regulation 3(4) **A young person shall not be employed unless there has been risk assessment carried out.**

The assessment shall take particular account of:

- *the inexperience, lack of awareness and immaturity of young persons;*
- *the fitting out and layout of the workplace;*
- *the nature, degree and duration of exposure to physical, biological and chemical agents;*
- *the form, range and use of work equipment;*
- *the organisation of processes and activities; and*
- *the extent of training provided.*

Regulation 19 **Every employer shall ensure that young workers are protected at work from risks to their health and safety.**

This regulation is supported by guidance 'Young People at Work' published by the HSE - ISBN 0 7176 1285 6.

No young person should be employed for work:

- *which is beyond his physical or psychological capacity;*
- *involving harmful exposure to toxic or carcinogenic agents;*
- *involving harmful exposure to radiation;*
- *involving risk of accidents due to lack of experience or training; and*
- *where there is risk from extreme cold or heat, noise or vibration.*

There are special requirements for the employment of children (under the age of 16 years), and their parents must be given copies of relevant risk assessments.

Some employers take children on work experience when they are 15 years old. Schools and colleges have responsibility for the children. Advice on how to manage their safety is included in the HSE publication 'Managing health and safety on work experience – a guide for organisers' HSG 1999.

Pregnant women

There are some occupations and tasks that present risks to pregnant women. Employers must carry out and document a risk assessment when a woman informs them that she is pregnant, and it may be necessary to repeat the assessment during the pregnancy as the woman's health and capability change.

Regulation 16 **Requires employers to take particular account of risks to women of child bearing age, and for expectant, new and breastfeeding mothers when assessing risks from work activity when told in writing that a worker is pregnant.**

If risk cannot be avoided the employer must make changes to working conditions or hours, or offer suitable alternative work or give unpaid leave for as long as necessary to protect the health or safety of her or the child.

Regulation 17 **If a medical practitioner or midwife certifies that it is necessary for an expectant mother not to work at night for her health and safety, she must be suspended from work.**

Regulation 18 **The employer is not required to take any action until notified by new or expectant mother in writing.**

Guidance on this provision has been published by the HSE in 'New and Expectant mothers at Work' ISBN 0-7176-0826-3. [HS(G)122].

In assessing risk to your employees you must pay attention to workers who are new or expectant mothers and to take action to ensure that they are not exposed to any significant risk. Physical, biological and chemical agents, plus working conditions, may affect their health and safety.

6 Hazardous Substances

THE CONTROL OF SUBSTANCES HAZARDOUS TO HEALTH REGULATIONS

It has always been recognised that one of the most common hazards found at work is the use, handling and storage of hazardous substances. The Control of Substances Hazardous to Health Regulations (COSHH) were first introduced in the late 1980s and have been revised many times since then.

The regulations and associated ACOP are a useful tool of management. These are set out in a simple step-by-step approach which will help to assess the risks, implement any measures needed to control exposure and establish good working practices.

There are many positive benefits for staff and the employer from carefully following through the requirements of COSHH.

Substances hazardous to health

Following the principles of risk assessment, the first point is to identify all the hazardous substances used in the workplace. It is good practice to systematically examine the workplace, including all store cupboards, workshops, cleaning stores etc., and list all the substances found. Those covered by the regulations are defined as:

1. Substances or mixtures of substances classified as dangerous to health under the Chemicals (Hazard, Information and Packaging for Supply) Regulations (CHIP).

These can be identified by their warning label and the supplier must provide a safety data sheet for them. Many commonly used dangerous substances are listed in the HSE publication *Approved Supply List*.

2. A substance for which there is a Workplace Exposure Limit (WEL). These are stated on data sheets and in the HSE publication EH40. The WEL must not be exceeded.

3. A biological agent. This includes micro-organisms, parasites etc.

4. Significant levels of dust. N.B., some dusts have a WEL.

5. Any other substance which when present in a workplace creates a risk to health due to its chemical or toxicological properties or the way it is used, for example, gases which when present in high concentrations in air act as asphyxiants.

6. Carcinogens used in the workplace.

What is <u>not</u> a substance hazardous to health?

COSHH applies to virtually all substances hazardous to health **except**:

- asbestos and lead, which have their own regulations;
- substances which are hazardous only because they are:
  - radioactive;
  - simple asphyxiates;
  - at high pressure;
  - at extreme temperatures;
  - have explosive or flammable properties; and
  - biological agents if they are **not** directly connected with the work and they are outside the employer's control.

For the vast majority of commercial chemicals, the presence (or not) of a warning label will indicate whether COSHH is relevant. For example, there is no warning label on ordinary household washing-up liquid, so if it is used at work it is not considered under COSHH; but there is a warning on household bleach, and so COSHH does apply to its use in the workplace.

Routes of entry

- **Inhalation** – of dust, gas, mist, fume, vapour. This is the most common route of entry into the body.
- **Ingestion** – substances which are swallowed and will pass into the body by absorption
- **Injection** – a forceful breach of the skin, perhaps caused by injury. This can carry substances through the skin barrier.
- **Absorption** – passing through the skin into underlying tissue or bloodstream. This route is normally linked to occupational dermatitis.

Effects of hazardous substances

These include substances used directly in work activities (e.g. adhesives, cold disinfectants, paints, cleaning or bleaching agents). They can also be generated from work activities (e.g. fumes from soldering and welding, and dusts from cutting materials).

Examples of the effects of hazardous substances include:

- skin irritation or dermatitis as a result of skin contact, e.g. from wearing latex gloves
- asthma as a result of developing allergy to substances used at work, e.g. working with flour;
- losing consciousness as a result of being overcome by toxic fumes;
- cancer, which may appear long after exposure to the chemical that caused it;
- infection from bacteria and other micro-organisms (biological agents) e.g. legionella from water distribution systems, the collection, storage and disposal of clinical waste, spillage of body fluids or leptospirosis from rat urine.

Bacterial infection from rat urine (Leptospirosis).

WHAT COSHH REQUIRES

To comply with **COSHH** follow the seven steps:

1. ASSESS THE RISKS

Decide who is to carry out the risk assessment. The trained risk assessor should carry out a search of the department to **identify every substance** to ensure that no substance is left unaccounted for. Once the search has identified **all** substances, an inventory should be drawn up.

In assessing the risks these substances present to health, consider:

- How much of the substance is in use?
- How people could be exposed to it?
- Who could be exposed to it?
- How often is the exposure?
- Can it be absorbed through the skin?
- Can it be ingested (swallowed)?
- Can it be inhaled?
- Does it affect the eyes?
- Add to that information (much of it is available on the data sheet) any information on what precautions are currently taken.

Where there is a remaining unacceptable risk, decide on the action that needs to be taken to remove or reduce the risk to acceptable levels.

Check that control systems work effectively.

Consider whether the substance can be absorbed through the skin. Where this could occur, a biological monitoring programme may help to assess the risks.

If it is decided that there is no risk to health or the risk is trivial, and the findings have been documented, the risk assessment is complete and no further action is required at this stage.

2. RECORD AND REVIEW THE ASSESSMENT

A record of the main findings of the assessment must be maintained in writing, unless they are so simple that they can easily be recalled and the conclusions explained at any time.

Enough information must be recorded to explain the decisions that have been taken about whether risks are significant and the need for any control measures to be implemented. Record the action staff and others need to take to ensure hazardous substances are adequately controlled.

The assessment should be a living document, which should be reviewed:

- at not less than five-yearly intervals;
- whenever there is reason to think it is no longer valid; and
- where there has been a significant change in the work.

The assessment should state when the next review is planned. Records are mainly for the benefit of the staff and employer and form part of the management system to protect health, but others may want to see them, e.g. HSE Inspectors, safety representatives and safety committees.

3. PREVENT OR ADEQUATELY CONTROL EXPOSURE

The first action required by the COSHH Regulations is to prevent exposure to substances hazardous to health, if it is reasonably practicable to do so. Consideration may be given to:

- changing the process or activity so that the hazardous substance is not needed or generated;
- replacing it with a safer alternative; and
- use it in a safer form, e.g. pellets instead of powder.

If prevention is not reasonably practicable, exposure must be adequately controlled using one or more of these measures:

- totally enclosing the process;
- partially enclosing it and using extraction equipment - local exhaust ventilation (LEV);
- providing good general ventilation;
- using systems of work and handling procedures *which minimise* the chances of hazardous materials spilling, leaking or otherwise escaping; and
- trying to reduce the number of employees exposed, or the duration of their exposure, but only *after* considering, and where possible using, the above measures.

If exposure cannot be adequately controlled by any of the above measures, provide PPE, e.g. face masks, respirators, protective clothing, as a means of control. The regulations only permit the use of PPE to achieve adequate control if other means of control cannot be used alone.

Adequate control

Under COSHH, 'adequate control' means reducing exposure to a level that most workers could be exposed to day after day at work, without adverse effects on their health. For a number of commonly used hazardous substances the Health and Safety Commission has assigned a workplace exposure limit (WEL) to help define adequate control.

If there is a current WEL for a substance, it will be stated on the data sheet. A list of all current WELs can be found in EH40 (HSE Publication).

Workplace Exposure Limits

A WEL is set at a level that is not likely to damage the health of workers exposed to it (based on current scientific knowledge).

For substances with a WEL, exposure should be reduced to comply with that WEL.

COSHH - Eight principles of control

- Design and operate processes and activities to minimise emission, release and spread of substances hazardous to health.
- Take into account all relevant routes of exposure - inhalation, skin absorption and ingestion when developing control measures.
- Control exposure by measures that are proportionate to the health risk.
- Choose the most effective and reliable control options which minimise the escape and spread of substances hazardous to health.
- Where adequate control of exposure cannot be achieved by other means, provide, in combination with other control measures, suitable personal protective equipment.
- Check and review regularly all elements of control measures for their continuing effectiveness.

- Inform and train all employees on the hazards and risks from the substances with which they work and the use of control measures developed to minimise the risks.
- Ensure that the introduction of control measures does not increase the overall risk to health and safety.

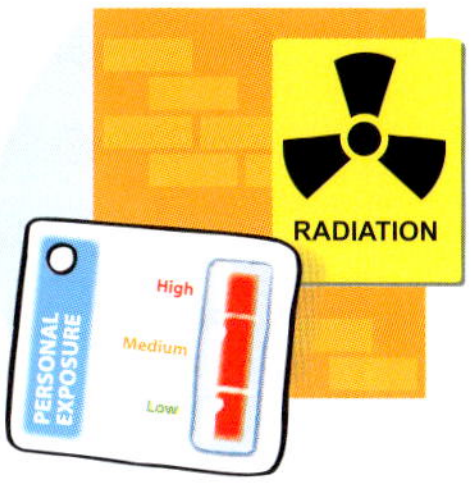

Skin absorption

Some substances can damage the skin itself while others can readily penetrate it, become absorbed into the body and cause harm. Consideration must be given to the need to protect the skin in deciding on control measures.

Carcinogens, asthma and genetics

Substances that can cause cancer, occupational asthma or damage to genes that can be passed from one generation to another must be controlled. Exposure to these substances must be reduced to as low a level as is reasonably practicable.

4. ENSURE THAT CONTROL MEASURES ARE USED AND MAINTAINED

Use the controls

COSHH requires all staff to make proper use of control measures and report any defects. It is the employer's responsibility to take all reasonable steps to ensure that they do so. This is the reason why staff must be given suitable training, information and appropriate supervision.

Maintain controls

COSHH places specific duties on the employer to ensure that the controls are kept in efficient working order and good repair. Engineering controls and respiratory protective equipment have to be examined and, where appropriate, tested at suitable intervals. The regulations require regular checks and thorough examinations for local exhaust ventilation equipment at a maximum of 14 months. Records of examinations and tests carried out (or a summary of them) must be retained for at least five years.

5. MONITOR EXPOSURE

Under COSHH, there is the requirement to measure the concentration of hazardous substances in the air breathed in by workers, where an assessment concludes that:

- there could be serious risks to health if control measures failed or deteriorated;
- WELs might be exceeded; and
- control measures might not be working properly.

Air monitoring must also be carried out when staff are exposed to certain substances and processes specified in Schedule 5 to the COSHH Regulations. Where it is appropriate to carry out personal air monitoring, the air to be sampled is the space around the worker's face from where the breath is taken, i.e. the breathing zone.

A record of any exposure monitoring carried out must be retained for at least five years if it measured environmental, general levels in the air. If the level of exposure to an individual was measured, records have to be retained for 40 years.

6. CARRY OUT APPROPRIATE HEALTH SURVEILLANCE

COSHH requires that health surveillance be carried out in the following circumstances:

- Where employees are exposed to a substance linked to a particular disease or adverse effect *and* there is a reasonable likelihood under the conditions of work of that disease or effect occurring *and* it is possible to detect the disease or health effect.

- Health surveillance might involve examination by a doctor or trained nurse. In some cases trained supervisors could, for example, check employees' skin for dermatitis, or check/ask questions about breathing difficulties where work involves substances known to cause asthma. A simple record of any health surveillance carried out must be kept. **COSHH requires that health records be kept for at least 40 years.**

Monitoring can also have a role in health surveillance.

7. ENSURE STAFF ARE PROPERLY INFORMED, TRAINED AND SUPERVISED

COSHH requires employers to provide staff with suitable information, instruction and training about:

- the nature of the substances they work with or are exposed to and the risks created by exposure to those substances; and
- the precautions they should take.

Employees must be given sufficient information and instructions on:

- control measures, their purpose and how to use them;
- how to use personal protective equipment and clothing provided;
- results of any exposure monitoring and health surveillance (without giving names); and
- emergency procedures.

These last steps are vital

It must be ensured that staff understand the risks from the hazardous substances they could be exposed to. Control measures will not be fully effective if staff do not know how to use them properly, deal with spillages or understand the importance of reporting faults.

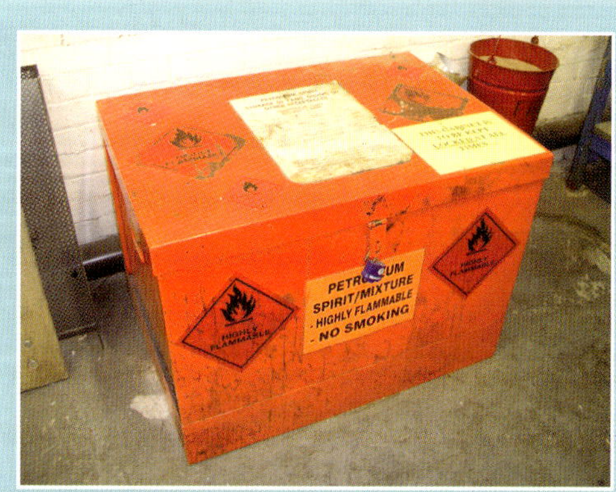

A reinforced storage box for chemicals, with sand bucket close by for emergency use.
(Image courtesy of Apple Tree Print)

7 The Workplace

If we look at the history of legislation designed to protect the health and safety of workers, we find that there have been basic requirements for the workplace since Victorian times. These have included safe access; adequate heating, lighting and ventilation and number of toilets etc.

The modern regulations are supported by an ACOP that gives details of how we can comply with the requirements.

THE WORKPLACE (HEALTH, SAFETY AND WELFARE) REGULATIONS 1992

Regulation 4 **Every employer shall ensure that every workplace shall comply with all requirements of these regulations.**

- *These regulations expand on the general duties under The Health and Safety at Work Act 1974. They are intended to protect the health and safety of everyone in the workplace and to ensure that adequate welfare facilities are provided for people at work.*

- *There is only one way of proving that the employer ensures that the workplace complies, and that is to have inspection records. If these show that the premises do not comply with the regulations, risk assessments should be carried out to allow actions to be prioritised.*

Regulation 5 **Equipment provided to comply with these regulations shall be maintained in good repair. Where appropriate, there shall be a suitable system of maintenance.**

- *Examples of equipment requiring a system of maintenance include emergency lighting, fencing, fixed equipment used for window cleaning, devices for limiting the opening of windows etc. This includes checks and planned maintenance.*

Regulation 6 **Effective and suitable provision shall be made to ensure that every workplace is ventilated by a sufficient quantity of fresh air.**

- *In many cases windows provide sufficient ventilation, although workers should not be subject to uncomfortable draughts. The ACOP gives details of the amount of air required, measured in air changes per hour or litres per minute. e.g. kitchens require more ventilation than on office.*

Regulation 7 **During working hours the temperature in the workplace shall be reasonable and a sufficient number of thermometers shall be provided to enable persons at work to determine the temperature.**

- *The temperature in a workroom should be reasonably comfortable without the need for special clothing. The temperature should normally be at least 16°C unless work involves severe physical effort when it should be at least 13°C.*

- *There is no maximum temperature defined in the ACOP, although risk assessments should be done if the temperature is excessive. Control measures include removing the source of heat, the taking of breaks, supplying cold drinking water etc.*

Regulation 8 **Every workplace shall have suitable and sufficient lighting, if practicable by natural light. Suitable and sufficient emergency lighting shall be provided.**

- *Adequate lighting should be provided to maintain safety and reduce eyestrain. Outdoor traffic routes should be lit after dark. Dazzle and glare should be avoided. Windows and skylights should be cleaned regularly to admit maximum natural light.*

- *Light is measured in lux, examples of levels are 100 lux in a store room and 400 lux in a room with detailed work taking place.*

- *Emergency lighting is required where sudden loss of light would pose a risk, e.g. with a loss of power. This should be regularly tested and maintained.*

Regulation 9 **Every workplace, furniture and fittings shall be kept clean.**
Waste material shall not be allowed to accumulate in a workplace except in suitable receptacles.

- *Interior walls, ceilings, furniture and work surfaces should be cleaned at suitable intervals, without exposing anyone to a health and safety risk. Arrangement should be made for collecting, storing and disposing of waste.*

Regulation 10 **Every room where persons work shall have sufficient floor area, height and unoccupied space for purposes of health safety, and welfare.**

- *The number of people allowed to work in a room depends on the size of the room and of the amount of space taken up by furniture, fittings, equipment and on the layout of the room.*

- *The total volume of the room divided by the number of people usually working there should be at least 11 cubic metres. This is a minimum and may be insufficient when furniture etc. is taken into account. Height above 3m cannot be taken into account when calculating the 11 cubic metres.*

Regulation 11 **Every workstation shall be arranged to be suitable for the worker and the task to be performed.**
A suitable seat and footrest must be provided if the task can be accomplished seated.

- *Workstations should be designed to be safe and comfortable, e.g. controls should be within easy reach etc. Examples include supermarket checkouts, hairdressers, assembly lines etc.*

Regulation 12 **Every floor in a workroom shall be of suitable construction, shall have no hole or be uneven or slippery.**

- *Floors and other surfaces shall be constructed and maintained to prevent slipping and falling, and arrangements shall be made for the minimisation of risks from snow and ice.*
- *Floors shall be kept free of obstruction that may present a hazard.*
- *As there are so many accidents and injuries caused by slips, trips and falls, the risk assessment for floors in all areas is important. Specific examples would include pedestrian paths, kitchen and bathroom floors, car parks etc. There is a considerable amount of advice available from the HSE on floor surfaces.*

Regulation 13 **Suitable and sufficient measures should be taken to prevent any person falling a sufficient distance to cause personal injury or be struck by a falling object.**

- *The consequence of falling from heights is so serious that a high standard of protection is necessary. Secure fencing is needed which is adequate to prevent objects falling.*
- *Secure fencing should be provided where there is a danger of falling more than 2 metres. It should be sufficiently high and filled in to prevent falls of people or objects over or through the fencing. Where necessary, upstands should be provided to prevent materials falling from the edge.*
- *Secure fencing shall be provided wherever possible even if there is no general access to the site or roof.*

Regulation 14 **Every window or translucent or transparent surface in a wall or partition or door or gate shall where necessary be of safety material or be protected against breakage, and be marked to make it apparent.**

- *Safety material shall be used if the glazed area is more than 250mm wide, and in doors or gates if any part of the panel is below shoulder level, and in windows, walls and partitions below waist level.*
- *Safety materials are materials that are inherently robust (polycarbonates etc.) and glass which, if it breaks, breaks safely. Premises should be checked for non-compliance with this regulation, and if necessary glass should be replaced or covered with a safety film. Protection could include a suitable barrier. All such glazing must be marked to reduce the risk of someone walking into it.*

Fire doors fitted with spots for visibility.

Regulation 15 **No openable window shall be likely to be opened, adjusted or closed in such a manner which exposes any person to risk to his health and safety.**

- *All windows, skylights etc. should be able to be adjusted, opened and closed without risk. Open windows should not project into an area where it is possible to collide with them, e.g. on footpaths.*

- *There have been many prosecutions where someone has fallen out of a window. Where there is a danger of someone falling, restrictors should be fitted and they should be routinely checked.*

Regulation 16 All windows and skylights shall be of design and construction that they may be cleaned easily.

- *Suitable provision should be made for cleaning of all windows and skylights in a safe manner. This could be a hardstanding for a ladder, but there are many types of access equipment including the provision of windows that are cleanable from the inside of the building with suitable harness points fitted.*

Regulation 17 Every workplace shall be organised in such a way that pedestrians and vehicles circulate in a safe manner.

Traffic routes shall be suitable for persons using them.

- *Traffic routes are defined as 'a route for pedestrian traffic, vehicles or both' and include stairs, ladders, doorway, gateway, loading bay or ramp. There are many accidents each year with pedestrians being in collision with vehicles. There should be separation of pedestrians and vehicles, particularly if the vehicle is reversing.*

- *Risk assessment will identify control measures. Sensible speed limits should be displayed along with warning signs for low headroom and blind bends. One way traffic systems may be necessary.*

Regulation 18 Doors and gates shall be suitably constructed.

- *There may be a need for a vision panel if the door or gate opens onto a corridor.*

Regulation 19 Escalators and moving walkways shall function safely and be fitted with any necessary safety devices.

- *There is HSE advice on escalator safety.*

Regulation 20 Suitable and sufficient sanitary conveniences shall be provided at readily accessible places.

Regulation 21 Suitable and sufficient washing facilities shall be provided at readily accessible places.

- *Minimum numbers of facilities are set, with the need for special provision for the disabled. Remote places and temporary worksites are included.*

Regulation 22 An adequate supply of drinking water and drinking vessels shall be provided for all persons at work.

- *Drinking water is usually supplied directly from the water main. Any water tank , cooler or cistern should be well covered, kept clean, tested and disinfected as necessary.*

Regulation 23 Suitable and sufficient accommodation shall be provided for any worker's own clothing that is not worn during working hours.

Regulation 24 Suitable and sufficient facilities shall be provided for any person at work to change clothes where the person has to wear special clothes for work.

- *For normal circumstances there should be a hook or peg provided for each worker to place outdoor clothing during the working day.*

- *If there is a need to wear special clothing e.g. catering workers, there shall be separate facilities for men and women. Facilities should be readily accessible, provide secure lockers for worker's own clothing and ensure the privacy of the user.*

Regulation 25 Suitable, accessible rest facilities shall be provided in one or more rest rooms or areas.

Facilities shall be provided for pregnant women or nursing mothers to rest.
Facilities shall be provided for persons at work to eat meals.

- *Rest facilities can be the staff restaurant if there is no charge to use it. If there is no facility to buy food, there should be provision to heat food and have a hot drink. Rest facilities should be no smoking at all times.*

- *Facilities for pregnant and nursing mothers to rest should be convenient for sanitary facilities and include the facility to lie down.*

8 The Safe Use of Work Equipment

Personal injury may occur to employees or other visitors to the site due to various hazards associated with the use of equipment. They include electrical shock, fire, and hazards due to faulty design or installation, mechanical hazards or noise.

In The Provision and Use of Work Equipment Regulations 1998, equipment is defined as 'Any machinery, appliance, apparatus, tool or installation for use at work (whether exclusively or not).' This is a very wide-ranging definition, including toolbox tools (hammers, screwdrivers etc.), photocopiers, hoists and slings, ladders, laboratory equipment, lathes, and drills etc.

By carrying out sensible risk assessments, the risks involved in using the many thousand pieces of equipment potentially used in an organisation can be controlled.

EQUIPMENT HAZARDS

These include:

- traps – for example power presses, paper guillotine, in-running nip between rollers;
- entanglement – hair and clothing is easily tangled in drive shafts and drills;
- ejection – equipment may throw off particles such as swarf, sparks from a grinding wheel or stones from a lawn strimmer; and
- contact – with equipment, sharp edges, hot surfaces etc.

Any control measures that need to be implemented must reduce the danger to the user, e.g. through the use of guards or safety devices. These must prevent as far as possible the operator or other persons coming into contact with the danger points of the equipment.

A shrink wrapping machine with various guards against trapping and heat injuries.
(Image courtesy of Apple Tree Print)

Printing press with various guards and emergency stop mechanisms.

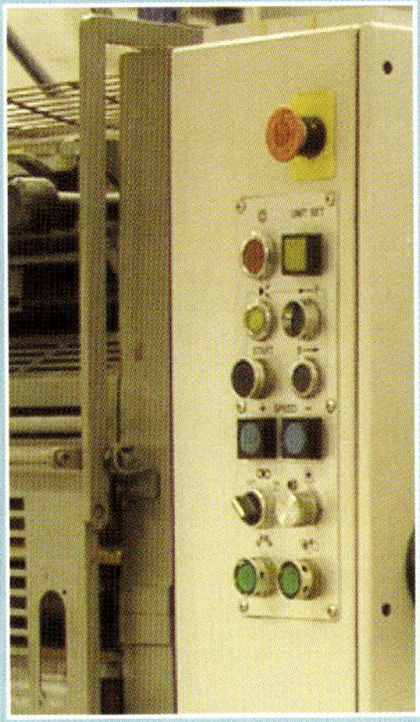

(Images courtesy of Apple Tree Print)

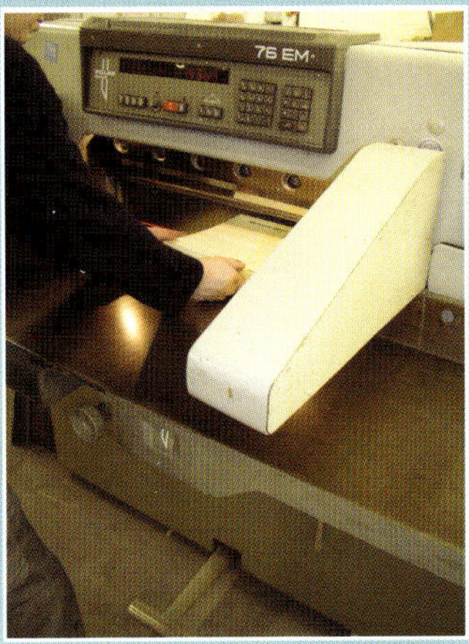

A guillotine with automatic light sensor guard.

GUARDING

There are five main types of machinery guard:

Fixed guard

A fixed guard has no moving parts and is placed so that contact with the machinery is impossible. It is often constructed of sheet or perforated metal, safety glass or polycarbonate sheet. It should not be removable without the use of a specific tool.

Interlocking guard

This guard works by controlling the machine and only allowing operation when in place, e.g. a microwave oven door is an interlock as the oven switches off when the door opens.

Automatic guards

This guard works with the machine, independently of the operator. The movement of the machine puts the guard in place before the operator is exposed to risk. Examples include power presses and guillotines.

Distance guards

This places the danger point out of reach of the operator, often with a funnel, grill or rail. An example is the filling entry to a food processor - the hand cannot be placed near the cutting bowl.

Adjustable guard

This is the least effective and reliable guard. It is adjusted by the operator before use, e.g. a circular saw or band saw.

THE PROVISION AND USE OF WORK EQUIPMENT REGS 1998 (PUWER)

These regulations amplify and make more explicit the general duties on employers to provide safe equipment at work. They came into operation on 5 December 1998, and replace the previous Provision and Use of Work Equipment Regulations (PUWER) of 1992 and many specific regulations, e.g. Woodworking Machines Regulations, Power Presses Regulations, Abrasive Wheels Regulations etc. The objective of the regulations is to ensure that work equipment should not result in health and safety risks, regardless of its age, condition or origin.

They were published with an approved code of practice and guidance, which gives information, explanation and advice on their implementation.

There are additional requirements for lifting equipment in The Lifting Operations and Lifting Equipment Regulations 1998 (LOLER) and supporting ACOP and Guidance.

Safe use of ladders.
(Image courtesy of Apple Tree Print)

Regulation 2 Defines the terms 'use' and 'work equipment'.

- ***Use***. *The definition of use is wide and includes all activities involving working with equipment such as starting, stopping, cleaning, repairing, modifying and servicing.*
- ***Work equipment*** *is also extremely wide, including all machines and tools.*
  Examples include computers, visual aid equipment, photocopiers, guillotines etc. It also covers toolbox tools, machines, apparatus, lifting equipment, ladders, etc. It does not cover privately owned motor vehicles.

Regulation 3 These regulations apply to equipment provided for use or used by employees at work.

- *There is already, under The Health and Safety at Work Act 1974, the general duty to provide and maintain safe work equipment. These regulations place duties on all employers to ensure that equipment is suitable and safe.*

Regulation 4 Every employer shall ensure that work equipment is so constructed as to be suitable for the purpose intended.

- *This regulation is the heart of this piece of legislation. It addresses the three important aspects of equipment:*
  a) its initial integrity;
  b) the place where it will be used; and
  c) the purpose for which it will be used.

- *The risk assessment carried out under Regulation 3 of The Management of Health and Safety at Work Regulations 1999 will help employers select equipment and assess suitability for particular tasks.*

- *Equipment must be suitable by design, construction or adaptation for the work it is provided to do in the particular location.*

- *When selecting work equipment, employers should take account of ergonomic risks. They should ensure that the equipment is installed, located and used to reduce risk to users and others.*

- *Equipment should be suitable for each process and the conditions under which it will be used.*

Regulation 5 **Every employer shall ensure that work equipment is maintained in efficient order and in a good state of repair.**

- *The word 'efficient' is used with regard to health and safety, and particularly items such as guards and fail-safe devices. Maintenance and checking should be done by competent persons and must be carried out in accordance with manufacturer's instructions. If inadequate maintenance could cause the equipment to fail in a dangerous way, a formal system of planned preventive maintenance may be necessary.*

Regulation 6 **The employer shall ensure that where the safety of equipment depends on installation conditions, it is inspected after installation but before first use. If conditions could cause deterioration, equipment should be inspected at reasonable intervals and after exceptional circumstances.**

- *Risk assessment should identify when inspection should be carried out. Inspection should include visual checks, functional checks and testing. Competent persons should decide on the nature of the testing and carry it out.*

- *Records must be kept and accompany the equipment.*

Regulation 7 **Where there is a specific risk, the employer shall ensure that use, repair, maintenance and servicing is restricted to authorised (trained) persons.**

- *Risks should be controlled in this order:*
  - *eliminate the risk;*
  - *physical measures (guards); and*
  - *safe systems of work.*

- *Only persons receiving information, instructions and training should be allowed to use, repair or maintain equipment. Where there are apparent significant risks, the employer should have a system of only authorised users having access to equipment, e.g. fork lift drivers having keys for the truck.*

Regulation 8 **Every employer shall ensure that every user of work equipment has received adequate health and safety information and, where appropriate, written instructions pertaining to the use of that equipment.**

- *This adds to the general duty to provide adequate information and instruction to ensure the health and safety of the user. Written instruction refers to the manufacturer or supplier's manuals and notices. This includes fixing and installation instructions. Information should be presented clearly in English and other languages as necessary. Sometimes employers find it helpful to prepare simple instructions in bullet points that can be attached to or close to equipment.*

Regulation 9 Every employer shall ensure that every user of work equipment has received adequate health and safety training including the risks in using the equipment and precautions necessary.

- *The obligation to train extends to users and to those that manage or supervise users, and should ensure that the user is competent – i.e. they are observed using the equipment correctly.*

- *Drivers of self-propelled equipment (tow trucks, lift trucks etc.) and chain saws* ***must*** *be trained and certified.*

Regulation 10 Every employer shall ensure that any item of work equipment purchased after January 1st 1993 complies with relevant UK legislation and EC directives.

- *This regulation aims to ensure that all equipment meets minimum requirements. Where appropriate, the equipment should display the 'CE' mark that confirms that it complies with The Machinery Directive and the Machinery (Safety) Regs 1992.*

- *Ensure that operating instructions and certificates of conformity are provided.*

Regulation 11 Every employer shall ensure that dangerous machinery is adequately guarded during use.

- *This regulation covers risks arising from most mechanical hazards and replaces existing legislation on machinery guarding. There is a wealth of advice and guidance published by the HSE and industry bodies on guarding of machinery; guidance deals with the general principles.*

- *The general risk assessment identifies the hazards of equipment, and the measures taken to control risk, including guarding.*

- *If guarding is necessary, there are four levels of measure that may be taken:*
  - *a) enclosing guards;*
  - *b) other guards or protection devices;*
  - *c) protection devices (push sticks etc.); and*
  - *d) information, instruction and supervision.*

- *The measures necessary must take into account the setting, cleaning, maintenance and repair of equipment.*

Regulations 12–30 These regulations cover the specific properties of certain types of equipment, including stop controls, emergency stops, isolation from energy source, controls, stability and lighting.

They also include specific regulations about passenger operated vehicles, mobile equipment etc.

More information on these regulations etc. is available from the HSE.

The ACOP also include the following provisions:

ACOP 32 You should ensure that work equipment is erected or dismantled in a safe way, in particular observe manufacturer's or supplier's instruction where they exist.

- *Safe systems of work (formal procedures) should be followed to achieve safety in erection and dismantling. It should be planned and hazards identified. The work should only be carried out by COMPETENT employees.*

ACOP 50 You should ensure that if there is a risk of work equipment being struck by lightning, appropriate safety measures are followed.

- *Assess the risks arising from the site and the height of the equipment, e.g. fairgrounds and cranes. Sometimes the protection is to stop work, e.g. golf professional in a lightning storm.*

LIFTING EQUIPMENT

Lifting equipment is defined as 'work equipment for lifting or lowering loads and includes its attachments used for anchoring, fixing or supporting it'. It includes any lifting accessories that attach the load to the machine.

Examples include:

- passenger lift
- rope and pulley
- dumb waiter
- lifting crane
- vehicle inspection hoist
- scissor lift
- ropes used for climbing
- hoists
- lorry fitted cranes
- refuse vehicle loading arm
- vehicle recovery equipment
- vehicle tail lift

THE LIFTING OPERATIONS AND LIFTING EQUIPMENT REGULATIONS 1998 (LOLER)

These regulations are supported by an Approved Code of Practice (ACOP) and guidance. They apply in all premises and work situations subject to the HSWA and build on PUWER. Those with responsibility for lifting equipment under these regulations must also comply with PUWER and The Management of Health and Safety at Work Regulations (MHSW).

The regulations require the employer to assess all risks to lifting equipment, including issues such as stability, and take specific steps to ensure safety.

These include:

- marking the maximum load on all lifting equipment, including slings, for example, maximum numbers of persons in passenger lifts;

- checking all equipment regularly. This is often carried out by users, and includes slings etc. Passenger lifts should be checked for door and alarm systems, level settlement etc.;
- routine servicing. Equipment should be subject to planned maintenance;
- regular testing. Engineer's tests should be carried out every 6 months if the equipment lifts people, twelve months if not. The records of all tests should be retained. If the examiner finds a serious fault, they have to inform the enforcing authority of that fault;
- all users should be given training on the safe use of the equipment and their competence tested. Records of training should be retained; and
- for passenger lifts, there should be competent people available to return the lift to safety in case of it breaking down between floors.

ELECTRICITY AT WORK

The hazards arising from electricity are twofold: firstly the risk of death or serious injury from contact with electricity, and secondly the risk of fire and explosion where electrical discharge could be the source of ignition. Injuries associated with electricity range from shock and burns to eye injuries caused by arc welding.

Effects of electricity on the body

The body is controlled by small electrical currents passing from the brain to the various muscles. If contact is made with an external source of electrical current, a reaction will occur in the body.

The scale of the reaction will depend on the size of the contact current and the path it will take through the body. For example, contact with the left hand exiting to earth by the right hand will result in the current passing through the chest. As a minimum of 8 pulses per second of direct current will contract a muscle and keep it contracted, it is easy to see that the 50HZ of domestic current can cause serious effects on the heart and lungs.

The following are given as a guide to the effects of current passing through the body: (1mA is one thousandth of an amp.)

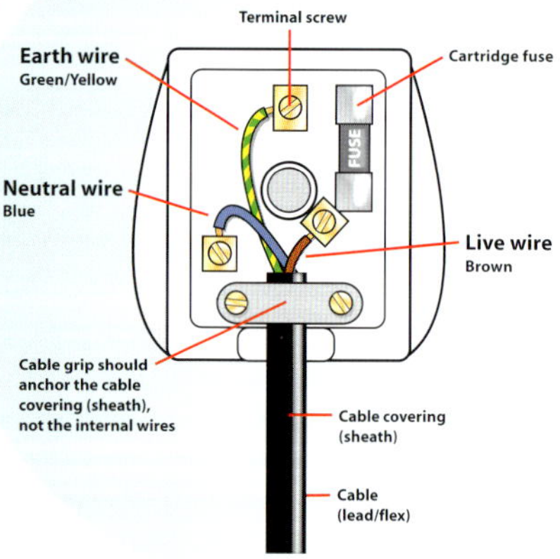

- 1mA ~ felt as a harmless tingle.
- 1-12 mA ~ more painful but contact can be broken as muscle control is retained.
- 12-20 mA ~ painful shock, causes panic and sweating, muscle control lost and unable to break contact. ('No let go intensity' is 16mA for men, less for women.)
- 20-50 mA ~ intense pain, paralysis of respiratory function.

Potentially dangerous wiring.

- >50mA ventricular fibrillation. The heart beats at 72 beats per minute and the current superimposed on heart rhythm disrupts timing. The heart flutters aimlessly and eventually stops. Injury and death can also be caused by damage to the nervous system and by burns.

Staff working with electrical supplies should be trained in the steps to take if someone receives an electric shock. This will include the need to switch off the current before attending to them and then take appropriate first aid action.

Precautions against electric shock

Most of these are common sense and include:

- good standards of installation and maintenance;
- use of double-insulated equipment and step-down transformers;
- regular testing and examination of equipment;
- use of earth leakage circuit breakers ELCB (also known as residual current devices RCD or residual current circuit breakers RCCB);
- use of PPE, barriers and lock off and permit to work systems by maintenance workers; and
- earth-bonding of metal equipment and work surfaces.

Appropriate signage to warn of danger of possible electric shock.
(Image courtesy of Apple Tree Print)

Electrical installation testing

The Electricity at Work Regulations require workplaces to have the electrical installation regularly tested by competent persons. This is usually completed by a competent person every five years, although some insurance companies require premises to be tested more frequently. The competent person provides a certificate which details actions that must be taken immediately, those that need to be taken as soon as possible and those that can be planned for the future.

Portable electrical appliances

The most simple definition of a portable appliance is 'one with a plug fitted'. The regulations require portable appliances to be:

- identifiable;
- checked regularly, usually by the user or during routine servicing; and
- tested regularly by a competent person.

User checks would include the following:

- Visual checks before use – is the equipment damaged? Evidence of cuts, burns or misuse? Is the plug damaged? Is the cable gripped securely into the plug? Should the user find any evidence of faults, the equipment should be removed from use and repaired by a competent person.

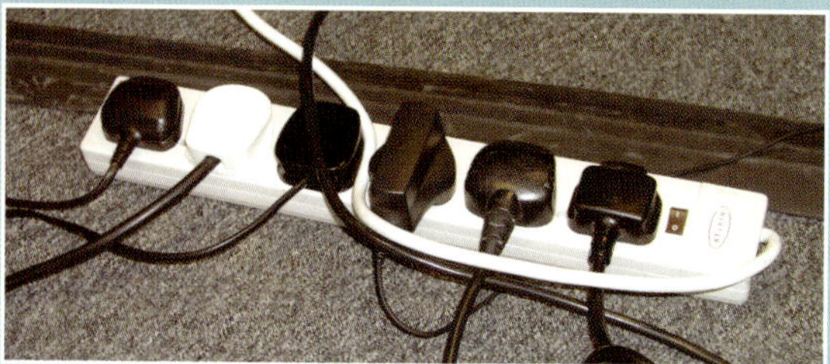

A trailing extension, with twisted overlapping cables, potential tripping hazard or electrical fire.

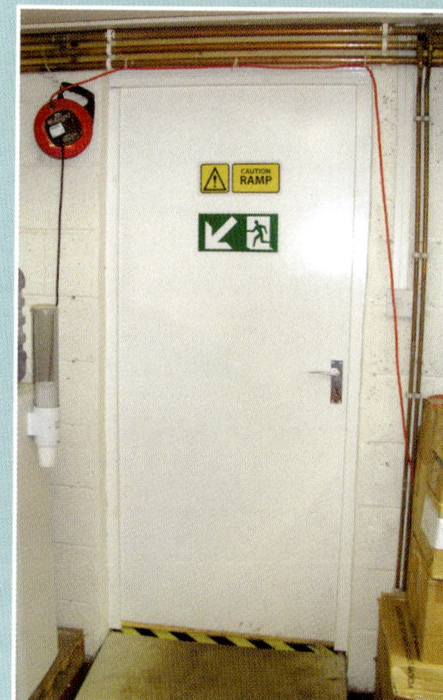

Inappropriate use of a cable extension reel.

- Testing (often known as PAT or portable appliance testing) should be carried out as often as necessary according to risk, e.g. an extension lead used by many people in many places will need to be tested more often than a computer that is installed and never moved. However, many employers have all electrical equipment tested every year or so to reduce cost and inconvenience.

NOISE AT WORK

Noise can be defined as unwanted sound, but if we are exposed to a loud enough noise for a sufficient period of time, it will affect our hearing. Noise can also cause stress.

Hearing

We hear sound through a complex mechanical process within our heads.

The OUTER EAR collects moving air, directing it towards the eardrum

The MIDDLE EAR magnifies the movement of the eardrum with three very small bones: the hammer, anvil and stirrup bones.

The INNER EAR consists of a jelly-filled sac called the cochlea. In this sac are hairs that are moved by the jelly reacting to the bones banging on the side of the sac. This stimulates a nerve signal that the brain interprets as sound.

Hearing loss

Hearing loss can be chronic or acute.

Chronic hearing loss, caused by long duration exposure can be:

- permanent threshold shift – where hearing is less effective, with low sound levels not picked up by the sufferer;
- noise-induced hearing loss – this is where the sufferer only loses the frequency of sound that they have been exposed to; and
- tinnitus – may become an irreversible, constant ringing in the ears, however, it can also be temporary and can wear off after around 24 hours.

Acute hearing loss, caused by short term exposure can be:

- acute trauma – where a specific exposure to a high level noise can damage

the ear drum, bones or cochlea. Often the cause has been gunfire or explosions; and
- temporary threshold shift – exposure to a loud noise can make the sufferer lose hearing for a short time, for example after attending a very loud concert.

Measuring noise

The unit of measurement of sound is the decibel (dB), on a scale of 1-160. A doubling of sound is represented by an increase of 3 decibels. Therefore, two loudspeakers producing a sound measured at 80dB each produce a combined sound of 83dB.

As the human ear does not react to different frequencies of sound in the same way, we measure a noise on a scale that is weighted to replicate the ear, and this acoustic scale is known as dB(A).

To measure noise exposure, we use noise meters that sample sound and record the noise level at regular intervals over a long period of time. To indicate a noise level equivalent to that we would be exposed to in a normal working day, an average is calculated by the equipment. This is known as the Equivalent Noise Level or Leq.

The Noise at Work Regulations

As from February 2006 the regulations require noise to be eliminated at source or reduced to a minimum. Wherever employees are or are likely to be exposed to risk from noise, an assessment must be carried out.

Two action levels are defined, together with actions that the employer must take.

- *First action level – daily personal noise exposure of 80dB(A) or peak 112 pascals.*
  At this level the employer must:
  - carry out noise assessments using competent staff;
  - provide instruction, information and training about risks, hearing protection etc.; and
  - provide hearing protection.

Appropriate use of PPE for hearing protection, in a noisy working environment.
(Image courtesy of Apple Tree Print)

- *Second action level - daily personal noise exposure of 85dB(A) or Peak action level of 140 pascals.*
  At these levels the employer must:
  - reduce noise exposure by means other than hearing protection where technically feasible;
  - establish signed hearing protection zones; and
  - provide hearing protection and ensure that it is worn;
  - provide health surveillance where risk is indicated; and
  - take the actions above where the risk justifies this.

The maximum exposure limit for noise is 87dB(A) or 200 pascals.

Noise control

This can be achieved in a variety of ways, including:

- engineering controls - purchasing quieter equipment or using a quieter process, using dampers or quiet mountings, operating rotating equipment as slowly as possible;
- location - moving the source away from work, or sometimes by turning the equipment around;
- enclosure - totally surrounding the machine in sound absorbing material;
- silencers - adding distance to pipes with expansion chambers and baffles etc. to reduce air noise;
- lagging - covering pipes etc.;
- damping - adding stiffeners;
- acoustic screens - noise-absorbing materials can be used to reduce reflection and echo;
- isolation - placing the worker in an acoustic room if there is no need to enter the noisy area for more than a short time; and
- hearing protection - always the last resort, can be ear plugs or ear muff type defenders.

9 Fire Precautions in the Workplace

Each year fire and the effects of fire cause substantial loss of life and extensive damage to property.

FIRE

To understand how to prevent fire, there is a need to understand the combustion process. There are three requirements for a fire to start and to continue to burn; these are referred to as the three sides of the fire triangle:

- *Fuel* – usually items containing carbon, such as wood, paper, liquified petroleum gas (LPG), petrol, living matter etc. Some inorganic elements are also combustible such as hydrogen and some metals such as magnesium.
- *Heat* – there has to be an initial source of ignition, such as friction, spark etc.
- *Oxygen* – this is usually present as approximately 20% of the atmosphere. It may also be generated from substances, such as oxidising agents. These are substances that contain oxygen, which is readily available in fire conditions, for example, hydrogen peroxide, sodium chlorate and nitric acid.

FIRE TRIANGLE

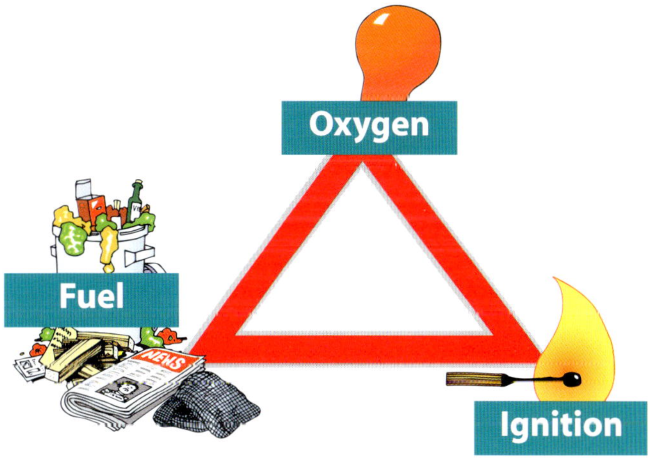

The causes of fire

These include electrical equipment, smoking, spontaneous ignition, sparks from machinery or equipment, open flames e.g. boilers and furnaces, hot work including welding, soldering, hot cutting and brazing.

FIRE SAFETY

There are two types of fire safety control measures: the building design and the management of people. The local fire authority will give specific advice on fire safety in your premises.

Emergency signs, fire points, fire escape and fire prevention signage.

THE BUILDING

(i) *Compartmentalisation*

Today, most buildings are designed to restrict the movement of fire and smoke. Each part of the building is separated from the next by walls, floors or doors that will prevent the spread of fire for at least 30 minutes. Each separate area is known as a 'zone'.

Areas with sources of ignition such as boilers, engines, cooking apparatus or electrical apparatus should be in separate zones as these are the areas where fires are most likely to start.

The zones are one of the major control measures in a building and it is important that they are understood and maintained. For example, one open fire door in a corridor could lead to fire and smoke spreading quickly through an entire building.

(ii) Escape routes

The design of the building should include adequate routes of escape.

Escape routes should be short and lead people to a place of safety, such as the open air or to a part of the workplace where they are not in immediate danger. People should be able to reach the open air without re-entering the workplace. Fire doors should be self-closing (fire doors to cupboards can be simply latched or locked shut).

Because stair openings allow fires to spread quickly, escape from the upper parts of some workplaces may be difficult. Therefore, most stairways will need to be separated from the workplace by fire-resisting partitions and self-closing fire doors. Escape routes should be clearly labelled.

(iii) Fire detection and warning

All workplaces should have arrangements for detecting and giving warning of fire.

In most cases an electrically-operated fire alarm system should be installed with manual alarm points sited next to exit doors and sufficient sounders to be clearly audible throughout the workplace. Automatic fire detection linked into an electrical fire alarm system should be considered where there may be some delay in a fire being detected by people in the workplace.

(iv) Fire-fighting equipment

Equipment has to be provided and checked regularly, and some staff trained to use it. But it is recognised that fire extinguishers are only of limited use in a large fire, and it is important that staff understand that they are only to be used if there is no risk to the user or others.

What's in it: WATER
Type of Fire: Wood, paper, fabrics

What's in it: FOAM
Type of Fire: General fires, flammable liquids

What's in it: POWDER
Type of Fire: General fires, plastics

What's in it: CO_2
Type of Fire: Electrical

What's in it: Fire Blanket
Type of Fire: Oil
Use: smothering flames

(iv) Maintenance and refurbishment

Sources of heat or combustible materials may be introduced into the workplace during periods of maintenance or refurbishment. Where the work involves the introduction of heat, such as welding or roof repairs, this should be carefully controlled.

THE MANAGEMENT OF PEOPLE

(i) Planning

Each workplace and area should have a fire plan including the action to be taken by staff if they discover a fire or if the alarm sounds. This should be developed by consultation with the staff involved and take into account the findings of the risk assessment.

For small workplaces this could take the form of a simple fire action notice posted in positions where staff can read it and become familiar with it.

Larger workplaces will need more detailed plans with notices giving clear and concise instructions of the routine to be followed in case of fire prominently displayed. The notice should include the method of raising an alarm in the case of fire and the location of an assembly point to which staff escaping from the workplace should report.

In order to assist disabled or sensory-impaired people to escape from fire other staff may need to be trained in the correct procedures.

In workplaces employing large numbers of employees, it may be appropriate to nominate 'fire marshals'. Their role might include ensuring that an area is completely evacuated during a fire evacuation, and reporting that fact to a control point. Others may have the task of closing down processes during an evacuation or ensuring that security is maintained whilst the workplace is evacuated.

The plan should be part of the health and safety procedures for the company. It should be included in all induction training and be part of the system for allowing visitors on site.

(ii) Training and instruction

All employees should be aware of the risk of fire and the action to be taken in case of fire in their area including:

- how to warn others of the fire, including the operation of the fire-warning apparatus provided;
- the location and use of escape routes;
- assisting or directing visitors or members of the public from the workplace;

- the location of a nominated assembly point;
- the use of the fire equipment provided; and
- how to summon the fire service.

There are two types of fire safety training:

- One type of fire lecture covers matters such as how fire starts and spreads and the effects of smoke and toxic gases. It will also cover typical fire situations, the use of extinguishers and other action to be taken. This is often held annually.
- The other type is held at the specific work area and includes the action to be taken on discovering a fire and what to do if the alarm sounds. This training is vital and needs to be repeated regularly during the year.
- Where fire marshals have been appointed, the training should ensure that their specific tasks are carried out efficiently and safely.

(iii) Fire drills

It is usual as part of the fire plan to carry out fire drills from time to time. A typical fire drill involves the sounding of the alarm and allowing the full evacuation of the building to occur. There will then be a debrief to find out if the fire plan was followed correctly. If the plan has not been followed, corrective action has to be taken, including coaching or retraining.

In some buildings it is very difficult to carry out a full fire drill, e.g. in a nursing home. In these cases it is important that the fire plan is checked in a different way, including desk top exercises or a fire drill using volunteers as residents. The fire authority will give advice on fire drills in specific premises.

False alarms occur from time to time and these in reality become fire drills. After each false alarm, also hold a debrief to see if the plan was followed.

THE FIRE PRECAUTIONS (WORKPLACE) REGULATIONS 1997

If you are in any doubt about the need to provide particular fire precaution measures you can check with the fire authority who will advise you. (The telephone number for non-emergency calls is in the telephone book.)

The regulations provide minimum fire safety standards in workplaces and employers are under a personal duty to comply with them.

The employer *must:*

1. assess the fire risks in the workplace;
2. check that a fire can be detected in a reasonable time and that people can be warned;
3. check that people who may be in the building can get out or to a place of safety;
4. provide reasonable fire-fighting equipment;
5. check that those in the building know what to do if there is a fire; and
6. check and maintain fire safety equipment.

Risk assessment

The fire precautions that are required depend on the risks and risk depends on current measures taken. Consider who might be hurt and how, where fires are likely to start, how they can start, where combustible materials are stored and used, the movement of smoke around the building etc. A competent and responsible person is required for each building to carry out the risk assessment.

Review your assessment from time to time, particularly if there is a significant change to your workplace or working practices, or you have frequent changes of staff.

Fire detection and warning

Check that fires can be readily detected and that staff can be warned promptly. The fire alarm system in a small office premises may be very simple, with perhaps just a bell, but in complex buildings it is likely to be an electronic detection and alarm system.

The fire panel must only be silenced or reset by trained and competent people.

Means of escape in case of fire

Staff and anyone on site should be able to get out or to a place of safety quickly and safely in the event of a fire.

Emergency routes and exits should:

- lead as directly as possible to the open air away from the workplace or to a safe area;
- be adequate for the type of workplace and the people likely to be in it; and
- be able to be quickly and safely used in the event of a power failure.

Evacuation routes must be marked with green and white directional signage.

Fire extinguishers.

Escape route and emergency doors

Whilst the workplace is in use, doors on escape routes should not be locked or fastened. They must be easily and immediately openable from the inside.

Provision of fire-fighting equipment

There should be sufficient fire-fighting equipment of suitable type to tackle a small fire. The equipment could consist of wall-mounted extinguishers, fire blankets or hoses. The fire risk assessment will direct which type of fire-fighting equipment is suitable. Advice can be sought from the fire authority or other competent person, when in doubt as to which fire-fighting equipment is required.

Planning for an emergency and training staff

Every employee must know what to do if there is a fire and be given appropriate information and training. In small workplaces this can be very simple, but in larger workplaces there must be a plan to ensure that everyone can leave the building quickly and that the emergency services are called promptly.

There is a duty on employees to cooperate with employers' directions and not to do anything that would place themselves or others at risk.

Fire action signage.

Maintenance and testing of fire safety equipment

Fire safety equipment must be regularly checked and maintained, including fire extinguishers, means of detecting a fire and warning alarm in case of fire, means of escape, and emergency lighting. Typically, this includes daily checks of fire escape routes, weekly checks of alarm sounders and annual checks of emergency lighting and extinguishers.

Routine inspections of fire safety equipment.

ENFORCEMENT

As with other inspectors, the fire authority inspector will firstly give advice, but if the inspector considers staff might be at **serious risk** in the event of fire, they can issue a notice requiring improvement to the fire precautions. Such a notice is known as an Enforcement Notice and failure to comply with it is a criminal offence. A reasonable amount of time must be given to comply with the notice and an appeal can be made (within 21 days) against the notice to a magistrates' court.

In very serious cases, which pose a **serious threat to life**, the fire authority can immediately serve a notice to prohibit or restrict the use of the workplace until the risk has been reduced. Failure to comply with the notice is a criminal offence. There is an appeal route against this notice through the Magistrates' Court, but it will remain in force until such time as the court says otherwise or the workplace has been made safe.

A deliberate or reckless failure to comply with the regulations, which places one or more workers at serious risk is in itself a criminal offence.

10 Health at Work

It is now universally accepted that work affects health. HSE statistics for 2001/02 claim that 2.3 million people in Great Britain suffer from an illness caused or made worse by work. Musculo-skeletal disorders are the most common work-related illnesses with 1,126,000 people affected. Over 500,000 people were affected by stress or depression, and breathing problems affected 168,000. In total, this added up to 32.9 million working days lost in 2001/02.

Many employers use an occupational health service to provide them with advice and assistance in this important area.

THE PURPOSE OF AN OCCUPATIONAL HEALTH SERVICE

To promote and maintain the physical, mental and social well-being of all staff.
(World Health Organisation)

The functions of an occupational health service

- assessing the health requirements of particular jobs through knowledge of the workplace and health hazards/risks;
- pre-employment health assessment;
- health surveillance of employees - looking for signs of ill health caused by hazards at work;
- management of the rehabilitation process;
- providing assistance in the risk assessment process;
- promotion of compliance with health and safety legislation, in liaison with health and safety managers and health promotion services; and
- analysing data on sickness absences to identify possible problems at an early stage (e.g. stress-related illness).

Health assessment on recruitment

Pre-employment health assessment ensures that applicants are fit for and placed in appropriate work. This usually requires a medical questionnaire to be completed by the applicant. Accurate assessment depends on the occupational health practitioner being informed of the work the new employee is expected to carry out. Although the medical details of the employee need not be reported, the employer must be satisfied that the employee is fit to do the job.

If there is a problem, occupational health staff will advise on changes to the job or environment that will allow the employee to do the job safely.

Health surveillance of in-service employees

Occupational health staff often fulfil the various health surveillance duties under health and safety legislation.

This may include examinations after injuries at work, after absence from work, fitness for work, surveillance for specific jobs and tasks (e.g. working with hazardous substances) or dealing with ill-health retirement.

Health promotion and education in the workplace

Occupational health staff often contribute to increasing the effectiveness of the organisation by enhancing staff performance and morale through reducing risks at work which lead to ill health, absence and accidents.

The Ethics of the Occupational Health Service

It is important that employers understand the professional values of the occupational health service, particularly the confidentiality of the medical records.

FIRST AID

First aid is defined as:

- treatment for preserving life;
- treatment for minimising the consequence of injury; and
- the treatment of minor injuries.

The scale and type of provision necessary in a workplace is covered by the First Aid Regulations and Approved Code of Practice.

The Health and Safety (First Aid) Regulations 1981

Regulation 3

An employer shall provide equipment and facilities adequate and appropriate for enabling first-aid to be given to employees if they are injured or become ill at work.

Employers have to provide facilities based on the risk within their premises, and must therefore carry out a risk assessment.

In assessing needs, the employer should consider:

- workplace hazards and risks;
- the size of the organisation;
- the history of accidents;
- nature and distribution of the workforce;
- remoteness from medical services;
- needs of travelling and lone workers; and
- annual leave and absence of first aiders and appointed persons.

These regulations do not oblige employers to provide first aid for any one other than employees. However, employers may wish to include the public in their assessment of needs.

First aid materials

The employer should provide materials and facilities needed to ensure that first aid is available at all times. The minimum level of first aid equipment is a suitably stocked and identified first aid container.

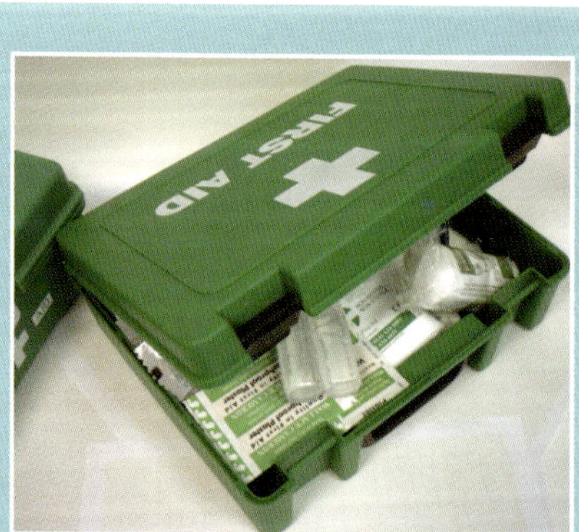

First aid box

There is no mandatory list of items to be placed in a container, but it would normally include a leaflet on first aid, sterile adhesive dressings, eye pads, triangular bandages, safety pins, wound dressings and disposable gloves. All containers should be identified by a white cross on green background.

First aid personnel

A first aider should hold a certificate of competence issued by an organisation approved by HSE. There are general competencies listed in the ACOP, but some first aiders may need extra training to suit their place of work. The certificates are time limited (three years) and refresher training and retesting is carried out before the certificate expires.

If the employers risk assessment requires first aid personnel, they should be provided in sufficient numbers and at appropriate locations to enable first aid to be given without delay.

There are no rules to decide how many first aiders to have. For example a low risk site may not need a first aider but an appointed person.

The selection of first aiders depends on factors such as reliability, communication skills, ability to cope with stressful situations etc.

Records should be kept of all first aid incidents and this is to be used as reference material during the assessment.

Appointed persons

Appointed persons signage.

If a first aider is not necessary, the minimum requirement is for an appointed person to be available to take charge of an incident.

Appointed persons should not attempt to give first aid for which they have not been trained, but employers are strongly advised to provide training on the following topics:

- what to do in an emergency;
- cardio-pulmonary resuscitation;
- first aid for unconscious casualties; and
- first aid for the wounded or bleeding.

MENTAL HEALTH AT WORK

Stress, depression and anxiety led to 13.4 million working days lost in 2001/02.

Work-related stress is defined in HSE guidance as the adverse reaction people have to excessive pressures or other types of demand placed on them.

This definition highlights the distinction between the beneficial effects of reasonable pressure and challenge and work-related stress.

Work-related stress is not an illness, but can lead to increased health problems if prolonged or particularly intense. These can include:

- heart disease;
- back pain, intestinal conditions and various minor illnesses; and
- psychological effects such as anxiety or depression.

Other behaviours detrimental to health that may result include skipping meals, drinking too much caffeine or alcohol and smoking cigarettes.

Signs of stress in an individual

Managers are often is a position to recognise stress in a colleague. This may be:

- changes in mood;
- changes in behaviour;
- deteriorating relationships;
- reduced performance;
- irritability;
- indecisiveness; and
- absenteeism.

Businesses can also suffer from the effects of stress, consequences of which may be:

- an increase in sickness absence – this can mean added work for other employees thereby possibly causing them stress or related illnesses;
- reduced staff morale;
- reduced staff performance; and
- staff seeking alternative employment, causing organisations the expense of recruiting, inducting and training replacements.

Legal requirements

There are legal requirements in relation to stress. Under The Health and Safety at Work etc. Act 1974, employers have a general duty to ensure the health of their staff, so far as is reasonably practicable. The Management of Health and Safety at Work Regulations 1999 require employers to carry out risk assessments.

Risk assessment is a means of approaching stress, in order to identify any problems with existing control measures that may be in place to prevent harm. The actions taken as a result of completing a risk assessment should reduce stress.

STRESS FACTORS

Factors that can contribute to work-related stress may be considered out of an employer's control and it is important to consider and acknowledge that staff may have domestic problems or are affected by events outside of work.

However, there are 'internal factors' that should be considered such as the employee's role, the demands on them including pressure or working long hours, or relationships within their working group. Other factors could be the culture of the organisation or management. External factors affecting the organisation, such as a competitor or an organisational change may also contribute to staff stress levels.

Action by employers

There have been a number of successful employee claims for work-related stress that have shown that employers must take action when a member of staff is diagnosed as suffering from the condition.

Employers should encourage managers to be understanding, and in particular listen to staff.

If stress is diagnosed managers must:

- maintain confidentiality;
- change the employee's job to prevent the condition becoming worse and allow sick leave until the employee and their GP are happy that return to work is appropriate;
- be flexible on return to work in terms of working times and the job;
- check others to ensure that no-one else is affected; and
- provide a confidential counselling service for employees to contact and act on general feedback from the service (although confidentiality must be preserved).

11 Ergonomics and Manual Handling

ERGONOMICS

Ergonomics can be defined as:

- the study of how a workplace and equipment can be best designed for comfort, safety, efficiency and productivity.

In other words, it is the study of people and their work. It is always about an individual and the relationship between the worker, the equipment they use and the environment they work in. Typically, ergonomics is involved in the design of workstations and safe systems of work when the job involves repetitive work, bending, stretching or twisting. Examples are:

- drivers;
- checkout operators;
- production line workers;
- hairdressers; and
- dentists etc.

Solving ergonomic problems involves the general principles of risk assessment.

If possible, remove the hazard. If that is not possible, carry out a risk assessment – with competent advice, to ensure the assessment covers the individual.

- control the risk to a tolerable level;
- mechanise the task;
- change or control the working environment; and
- redesign the work or equipment.

MANUAL HANDLING

The number of injuries and days off work caused by poor manual handling is detailed in the HSE statistics for 2001/02. These show that:

- musculo-skeletal disorders are suffered by 1,126,000 people in Great Britain;
- musculo-skeletal injuries caused 12.3 million working days to be lost.

Unsafe practice.

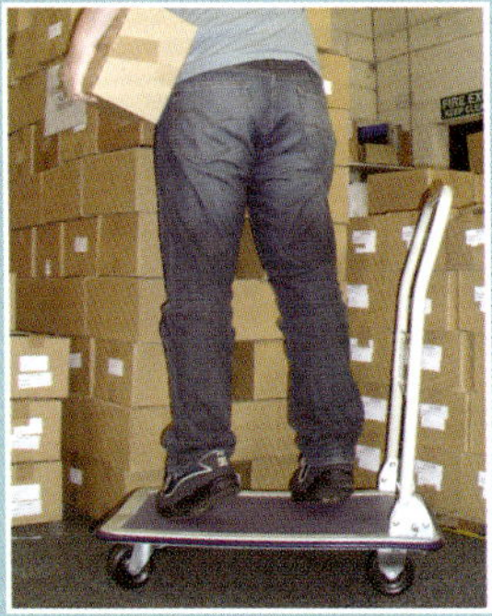
Unstable platform.

Incorrect posture.

In 2002/03 manual handling accidents caused 39% of all RIDDOR 3-day injuries.

Manual handling is the:

Movement or support of any load by physical effort, including:

- lifting;
- putting down;
- pushing;
- pulling;
- carrying; and
- moving.

It includes using mechanical aids such as trucks, trolleys etc.

Good manual handling.
(Image courtesy of Apple Tree Print)

Manual handling injuries include:

- fractures;
- damage to muscles, ligaments and tendons;
- spinal disc injuries;
- trapped nerves;
- abrasions and cuts;
- burns; and
- hernias.

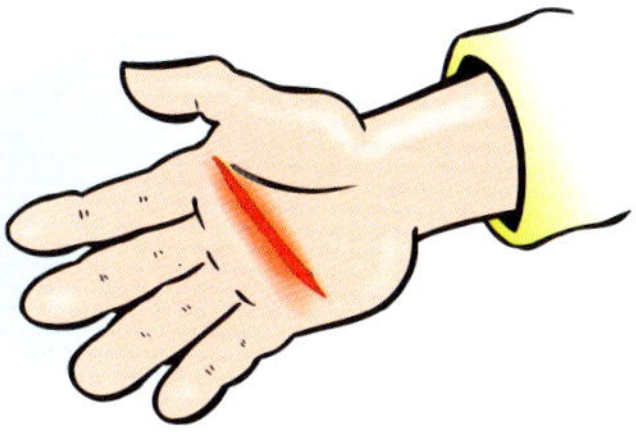

The most common injuries caused by manual handling are to the back.

The anatomy of the back is basically a stack of bones (the vertebrae) held in place by muscles and ligaments. Between the bones are discs of cartilage. Nerves enter the spinal cord through the gaps between the bones.

When the body twists, stretches or bends, the discs are put under pressure. If the disc moves out of line, it can be damaged or irritate one of the nerves, perhaps causing pain elsewhere in the body e.g. the legs. If a disc is damaged it may cause pain throughout the patient's life. The enormous pressure put onto back muscles sometimes causes strains and severe pain.

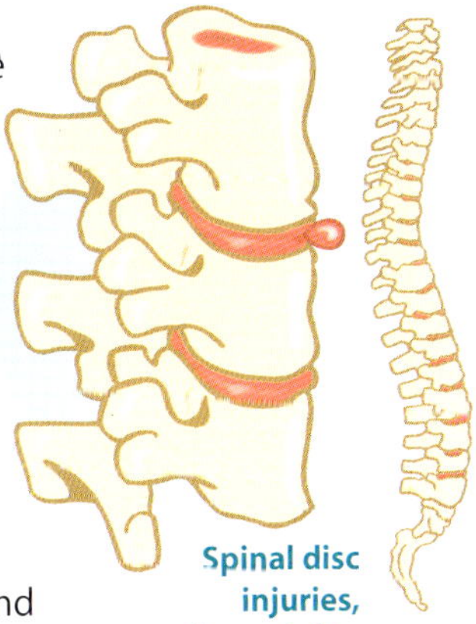

Spinal disc injuries, prolapsed disc

MANUAL HANDLING RISK ASSESSMENT

1. Avoid risk

The first responsibility under The Manual Handling Operations Regulations 1992 is to identify hazards and **avoid** manual handling risk. This requires a full review of all activities to identify manual handling tasks and then avoid them by mechanisation or automation.

2. Assess risk

If avoidance is not possible, risk assessment is necessary. This involves looking carefully at what is currently done to reduce risk and taking action to control risk to an acceptable level. The risk assessment must be carried out by competent staff and records kept.

The assessment must take into account the following:

TASK INDIVIDUAL LOAD ENVIRONMENT

The TASK

Look carefully at the job to be done. Does it involve handling away from the body? Movements such as bending, stretching or twisting? Is the movement over a long distance or is it repetitive?

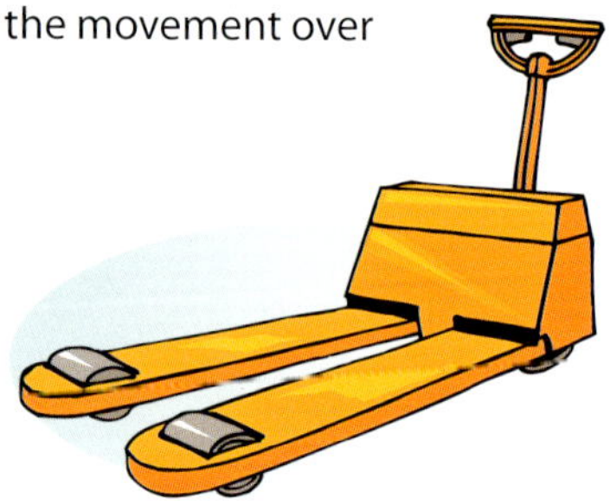

Control measures could include:

- using machinery/handling aids;
- improving the task layout;
- modifying the movement of the body;
- improving the work routine; and
- team handling.

The INDIVIDUAL

Consider the capability of the individual to carry out the task required. Are they fit and healthy? Have they been given information and training? Are they competent? Is adequate supervision provided? Is there any unusual requirement which would cause a risk to certain employees? (e.g. to pregnant women.)

Control measures could include:

- pre-employment medical;
- provision of information and training;
- ensuring that employees are competent in safe handling techniques; and
- providing supervision.

The LOAD

Look at the load. How heavy is it? Is it equally balanced or unwieldy? Is it stable? Does it have sharp edges? Will the contents shift?

Control measures could include:

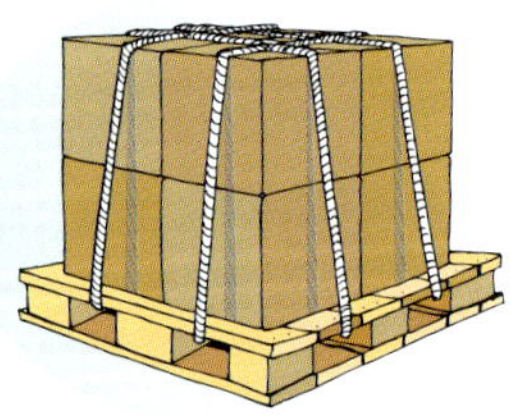

- making the load lighter by splitting it;
- making the load smaller;
- providing handles to make it easier to grasp; and
- making the load more stable.

The ENVIRONMENT

Consider the working environment where the operation takes place. Is there enough space to carry out the task? Is the floor even or slippery? Is there good lighting?

Control measures could include:

- improving workplace layout;
- improving floor condition;
- keeping to one level;
- improving environmental conditions; and
- good housekeeping.

SAFE LIFTING TECHNIQUES

STEP 1
Stop and think

Position of feet

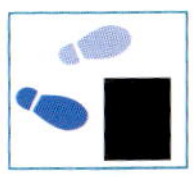

STEP 2
Position the feet

Position of feet

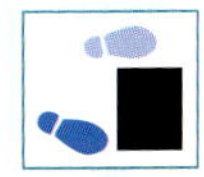

STEP 3
Bend the knees

Position of feet

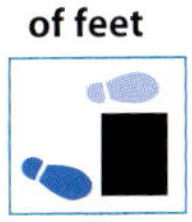

STEP 4
Get a firm grip, keep the back straight

Position of feet

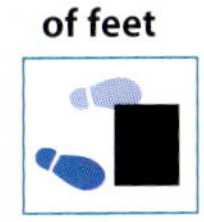

STEP 5
Raise with the legs

Position of feet

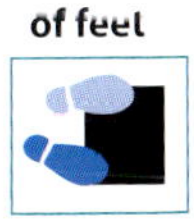

STEP 6
Keep the load close to the body

Position of feet

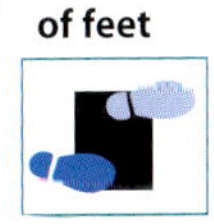

3. Provide information and training

Staff must be given sufficient information and training to implement control measures and keep records. It is important that a record of the ability of the employee to complete a task safely is retained. Training could include:

- how to avoid manual handling hazards;
- unavoidable and unfamiliar handling operations;
- use of handling aids;
- use of PPE e.g. gloves;
- hazards of the working environment;
- the importance of good housekeeping;
- an individual's capability; and
- good handling technique.

EMPLOYEES' RESPONSIBILITIES

Employees also have responsibilities – to co-operate by following procedures and to look after themselves and others.

SAFE USE OF DISPLAY SCREENS

There are accepted risks in the use of display screen equipment that are covered by The Health and Safety (Display Screen Equipment) Regulations 1992. The main requirement is for risk assessments to be carried out. The most common display screen is a computer and monitor, although other displays may be covered by the regulations, for example microfiche.

The hazards associated with the prolonged use of display screens are:

Work-related upper limb disorder (WRULD)

This is caused by the user sitting in the same position for many hours every day, carrying out repetitive tasks. Examples are injuries and pain to the fingers, hands, wrists and shoulders. Sometimes these injuries are known as repetitive strain injuries (RSI) and sometimes they are due to the posture adopted by the user.

Visual fatigue

This may be caused by prolonged activity at the screen causing the muscles in the eyes to tire. Every time the user moves their eyes from the screen, the eyes automatically refocus, using tiny muscles on the lens. This happens thousands of times a day, and the eyes tire. The use of a document holder has been shown to reduce visual fatigue.

General fatigue and stress

Neck, back and shoulder pains arising from the adoption of constrained postures at a keyboard. The use of display screens can be an isolated job, with workloads and targets to be met. This can be a cause of stress and fatigue.

Pregnancy

There are no reported risks to the mother or baby from the equipment, but as the woman becomes larger, musculo-skeletal problems may develop due to prolonged sitting.

USERS

The regulations only apply to 'users', as defined by the regulations. A rough description is that a 'user' habitually uses display screen equipment for a prolonged period of time every day and for a significant part of their work.

RISK ASSESSMENT

The assessment consists of a detailed inspection of the workplace and particular workstation. Four items are considered for the assessment:

- the equipment;
- job design;
- environment; and
- the user.

The standard we expect at a workstation for a user is as follows:

1. The chair must be on five castors and be adjustable in height and for the position of the backrest.
2. The user's feet must reach the floor or be supported by a footrest.
3. The monitor must be directly in front of the user with the top of the screen level with the eyes of the user. It should swivel and tilt to suit the needs of the user.
4. The keyboard must be movable, independent of the screen, and be in good repair and condition.
5. The screen should not be affected by glare. Bright lights can cause glare directly from light bulbs and strip lighting or indirectly by reflection from light coloured walls, glass or metal etc. Only as a last resort should anti-glare screens be used.
6. In addition, the user should plan their workload to include a break away from the machine every 50-60 minutes. This break should last 5 minutes or so. The break is not an excuse not to work, but takes the member of staff away from the screen to where other tasks may be performed, such as filing, telephoning etc. It is far better to have 5 minutes every hour than 20 minutes every two hours.
7. Users are entitled to eye and eyesight tests on request. The optician will recommend when the next test is required and if spectacles are necessary. If the user's current spectacles are not suitable, the employer has to pay for them.
8. Users should be encouraged to adjust furniture to allow the correct posture to be adopted when working on display screen equipment. The elbow should be at an angle of 90° with the lower arm parallel to the floor.
10. Each user should receive training about the hazards and potential health effects of using display screen equipment.

The assessment is not complete until the preventive and protective measures have been implemented. This may entail changes to furniture and the training of the user.

PORTABLE DISPLAY SCREENS

The requirements detailed above show that laptop computers and similar display screens cannot comply with the regulations. If they are used for prolonged periods of time, they should be adapted to meet the standard. This is easily achieved e.g. by using a separate keyboard and mouse and putting the laptop on a stand. The screen then acts as the monitor.

12 Health & Safety Training

All employees must be trained how to carry out their job safely and how to work in their workplace safely. Training involves the provision of information and how to implement that knowledge. The provision of information, instruction and training is contained in all modern legislation. The employer must provide this in a comprehensible way to employees, in particular on induction and in line with control measures required by risk assessments.

Training includes many different ways of communicating procedures and practices and of ensuring competence.

All training must be recorded with notes on how the training was carried out, that understanding was checked and how competence was assessed.

A training session using a PowerPoint™ presentation.

LEGAL REQUIREMENTS

HSWA – Section 2 requires the employer to provide information, instruction, training and supervision necessary to ensure the health, safety and welfare of employees.

MHSW Regulations – Regulation 13 requires employers to take into account the capability of the individual when allocating a task. To comply with this regulation the employer must be satisfied with the competence of the employee.

Employers should ensure that employees are provided with adequate health and safety training:

- on recruitment;
- when the job changes; and
- from time to time.

This training should take place during normal working hours.

In addition to these general requirements, most regulations now include a specific requirement to train staff. Examples include Manual Handling Operations Regulations, Control of Substances Hazardous to Health Regulations, Personal Protective Equipment Regulations etc.

PLANNING TRAINING

To ensure that it is provided for the right employee at the right time, there needs to be a systematic approach to the management of training.

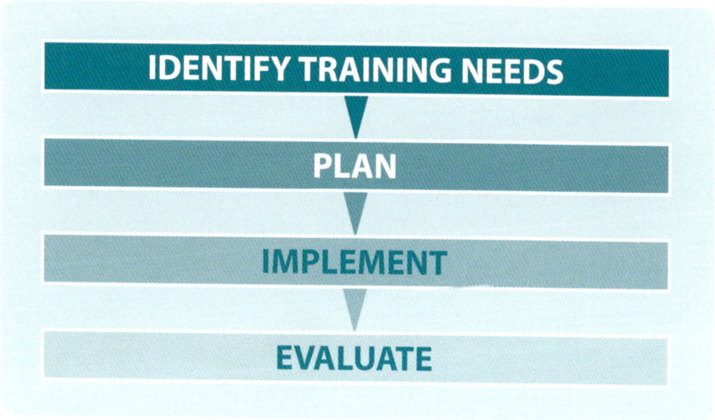

Identification of training needs

This needs to be carried out in a systematic manner, starting with each individual's job description. Supervisor's observations and discussions with the member of staff will also contribute. Training needs may also arise after an incident or with new legislation.

Planning training

We have all attended training that has been really helpful and interesting, and attended some that has been a waste of time and boring. When planning training, consider the needs of the trainees, the environment in which the training will take place and the trainer.

The *trainee* will need to be informed of when, where and why the training is taking place. It must be relevant to their current or future job, and be held at a suitable time. The ability of the trainee and their understanding of language must be taken into account. Some training will be one-to-one or in small groups, at other times it may be larger groups of staff.

The *environment* should be as comfortable as possible, with the correct facilities for the type of training taking place. Chairs, tables, visual aid equipment and refreshments may be needed. Try to eliminate unnecessary interruptions. The room should be warm, well-ventilated and well-lit.

The *trainer* needs to be competent in training, able to use a variety of training techniques and visual aids. They should have clear objectives and be able to meet them in the allotted time.

When the programme has been planned it needs to be adequately funded. There are direct costs, including the room, videos, paper, refreshments etc. Indirect cost is usually the time of the trainer and trainees to attend. To make training happen, it may be that extra staff have to cover for the trainees and this must be taken into account.

Implement training

There are many training techniques that can be used, from lectures and presentations to role play and coaching. The choice of technique may depend on the subject, but it is often said that the best learning takes place when trainees do something practical rather than just listen. Learning is certainly easier when the training is fun, and it is important that we make training events enjoyable rather than a burden.

Evaluation of Training

There are four basic levels of evaluating training:

- Reactive – this is carried out by asking the trainees questions, often with a simple form. Immediate feedback is received on matters such as the training environment, refreshments, the trainer etc.
- Knowledge – it is important that we find out if learning has taken place. This is achieved by testing knowledge and understanding with questions. These can be verbal or written, and a record should be retained.
- Competence- we must have evidence that an employee can do a job safely, and their competence must be observed at their workplace. Again, a record should be kept.
- Organisational – the most difficult evaluation of training is to look at the whole organisation some time after the training and measure improvements. You may be able to measure accident rates, number of claims, RIDDOR reports, costs of accidents etc. However, we must remember that other things may impact on these results and not just staff training.

Evaluation of training should result in training programmes and individual sessions becoming more effective.

TRAINING PROGRAMMES

If planned and resourced adequately, training will be on-going with a programme to be followed throughout the year. This will include the following:

Induction

Before starting work, all new employees should receive training in essential health and safety matters such as fire safety, manual handling, first aid, accident reporting etc.

Until the manager or supervisor is satisfied with the competence of the employee to carry out a task, there will need to be a higher than normal level of supervision to ensure the employee's safety.

Employee's should not be asked or expected to carry out any task until they have received suitable training.

Refresher training

From time to time, all employees should receive routine refresher training on health and safety aspects of their job. The frequency of this refresher training will depend on risk, for example, training on fire safety may need to repeated annually.

TRAINING RECORDS

The maintenance of suitable training records is very important. Training records allow managers to identify the training that each member of staff has received.

To defend a personal injury claim, it is often necessary to prove that:

- the individual attended a training session;
- the objectives of the session;
- that the trainee understood the training; and
- the trainee is competent to do that particular task.

The individual's training record should indicate who carried out the training and the training materials used. If the training was based on a skill, such as using equipment safely, the competencies will be based on the instruction manual and risk assessment.

13 Health & Safety Inspections

As part of an effective management system for health and safety at work, there should be a simple, documented policy that states commitment, defines responsibility (and the route of accountability) and sets out a clear strategy. This policy will also contain the systems for assessing risk and prioritising control measures.

Managers then need to be aware of how the strategy is progressing, what current risks they face and the effectiveness of the implementation of the systems in reducing risk. This is the role of performance measurement and inspection, and is summarised in the flowchart below.

HEALTH AND SAFETY ACCOUNTABILITY FLOWCHART

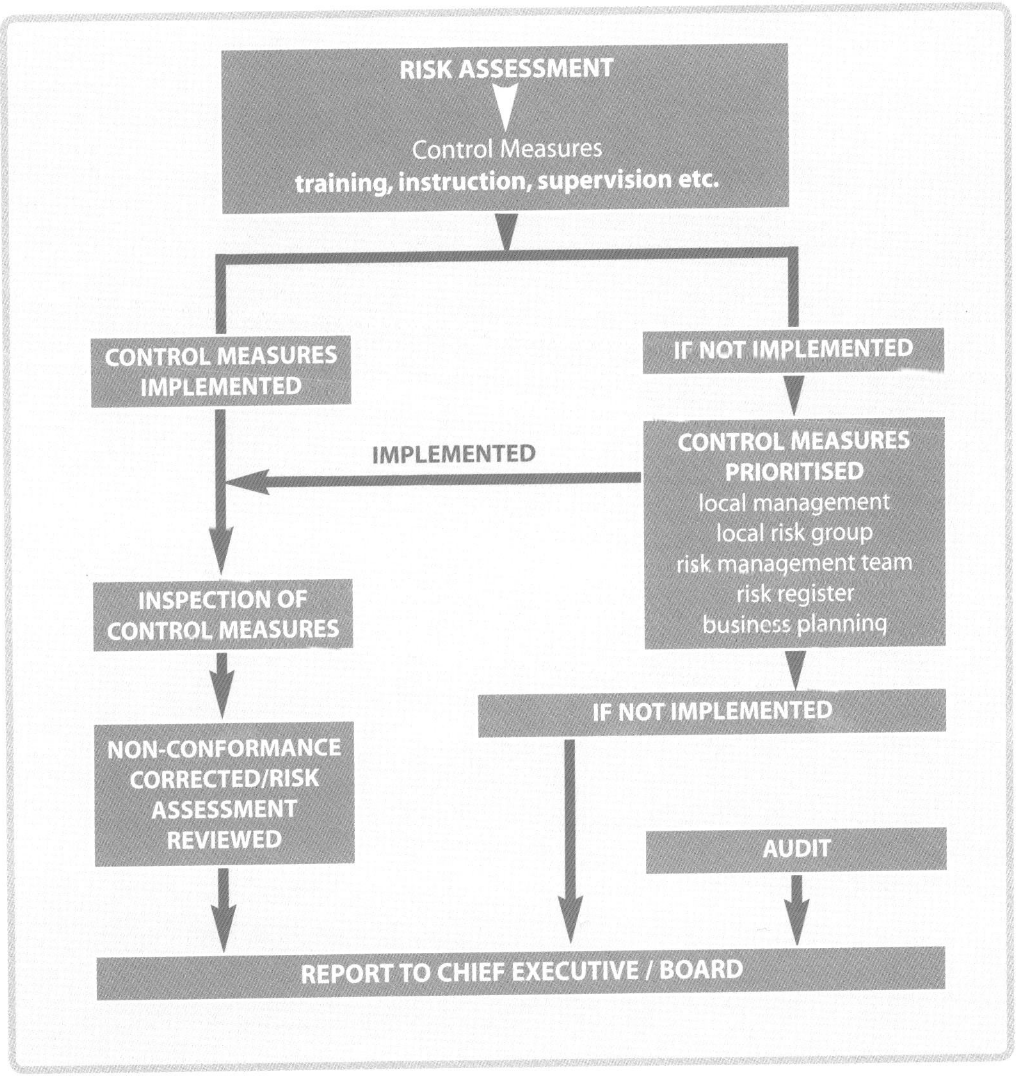

There are many types and levels of inspection, from a basic workplace check and hazard spotting exercise, through to a complex and detailed audit of the organisation. The objectives of carrying out such inspections include:

- to determine what needs to be done to improve health and safety;
- to confirm that the workplace conforms to legislative standards; and
- to keep senior management informed of progress.

TYPES OF WORKPLACE CHECK

Safety survey

This is a detailed examination of a narrow field of activity, and is often carried out prior to moving into new or refurbished premises or before a new work practice is adopted. It can also be helpful in checking the effectiveness of a safe working system, for example, electrical safety procedures, working at height and manual handling.

Typically, the survey includes details of administration, environmental conditions, operational safety and emergency procedures.

Safety survey.

Safety tours

This is an unscheduled tour of a workplace by a number of people. These often include managers, personnel officers, directors, estates managers etc.

The purpose is to ensure that particular control measures are being maintained, for example, fire protection or personal protective equipment. A safety committee may carry out a safety tour before a meeting to highlight certain issues for discussion.

Safety sampling

This is a systematic assessment of the workplace and is carried out by a number of people at the same time.

For example, once a week, at different times of the day, a number of supervisors will meet and be briefed on the day's sample. Examples of suitable practices to sample could be slips, trips and falls, manual handling or the use of hazardous substances. They then go to an area of the workplace, for say 15 minutes, and make a note of any potential incident using a checklist.

On their return they compare notes and decide what action, if any, is necessary from their findings. Over a period of many weeks and months, trends locally or organisationally can be noted and the effectiveness of systems evaluated.

Safety audits

An audit is a regular, systematic and critical look at all aspects of an organisation's activity, in this case in health and safety at work.

It is usually carried out by an experienced and competent health and safety professional, who will look at the whole management system. Many audits are based on Regulation 5 of The Management of Health and Safety at Work Regulations 1999, which require the employer to:

make and give effect to arrangements for planning, organisation, control, monitoring and review of protective and preventive measures.

Typically, an audit examines the delegation of responsibility, competence, training plans, risk assessment and prioritisation systems, emergency planning, incident statistics, etc. It may be carried out by examining all documentation, interviewing key personnel and observing activity.

The audit report is usually made to the owner or chief executive, and is a measure of the effectiveness of the health and safety management strategy. It is often said that an audit looks at the things that we do and how we do them, and an inspection looks at our compliance with those procedures.

Safety inspections

This is a detailed examination of a workplace and ensures that it meets *organisational* and *legislative* standards. It is important to know if control measures identified in risk assessments are consistently implemented and if the premises comply with the law.

THE PROCESS OF CONDUCTING HEALTH AND SAFETY INSPECTIONS

There are four stages in conducting a health and safety inspection:

a) planning and preparation;
b) performing the inspection;
c) reporting the findings; and
d) following up the action plans.

a) Planning and preparation

Firstly, make sure that everyone knows where and when the inspection is to take place.

Opinions are sometimes divided on whether or not we should give notice of an inspection. If you do, it is likely that there will be some clearing up and heightened awareness of health and safety issues up to and during the inspection. If you don't give notice, it is thought that you will see the workplace as it normally operates. In practice, within a few minutes of the start of the inspection, everyone knows what is going on and acts accordingly. If there has been some clearing up, it is often only superficial and easily recognised.

The usual advice is that everyone should know when the majority of inspections are to take place – it does mean that they are aware that you are coming!

Checklist design

The next part of preparation is to design a checklist. It is very difficult to carry out an inspection without a checklist - items are forgotten, observations and responses to questions are missed or not written down and not traceable when the report is written.

The opportunity should be taken to use all current documentation on procedures and practices, risk assessments and control measures and incident data to prepare a thorough and comprehensive checklist. For each element, such as use of hazardous substances, manual handling, workplace etc., we should include legislative and organisational requirements.

Checklists sometimes include scoring processes that may be weighted in terms of importance in terms of risk. This can then be used to give scores for each element or an overall score and indications of risk prioritisation.

An example for a scored inspection of manual handling could be:

CHECKLIST DESIGN

Scores

- 0 - nothing done
- 1 - some attempt at action, but not effective
- 2 - action partially effective, plenty of room for improvement
- 3 - implemented but not fully effective
- 4 - fully implemented and effective

MANUAL HANDLING	Score	EVIDENCE AVAILABLE
Is there a policy available for manual handling operations which sets out responsibilities for risk assessments and management including named assessors?		
Are all manual handling operations assessments reviewed regularly ?		
Are all hoists and slings marked, checked and tested regularly ?		
Have all staff received up-to-date training on the safe movement of loads ?		
Are staff training records complete and readily available ?		
Is manual handling operations performance audited and reported regularly ?		
TOTAL		

It is a simple process to then add weighting to each part of each element depending on importance in terms of your risk.

The checklist could include your own notes on questions you would like to ask, documents you would like to see etc. This ensures that you are not distracted and miss out any important items and the record of the inspection may be useful to justify your future recommendation.

Finally, you may wish to record on this checklist the numbers of reported incidents, RIDDOR notifications and sickness absence.

b) Performing the inspection

There are five important things to do during the inspection:

- be in control;
- ask questions;
- listen to the responses;
- watch, observe activity; and
- hold an exit meeting.

Always be in control of the inspection, see the people that you need to see, even if it is inconvenient, have a break when you want to and go where you need to.

It is helpful to have the questions that you need to ask, and the things you want to see detailed in the checklist prepared earlier.

- Questions – these need to be simple to understand and be phrased in such a way as to give the interviewee the opportunity to tell you some detail. The question should begin with words such as 'who, what, where, why, how, and when'. If we use words such as 'do, does, can' etc., the response could be 'yes' or 'no'.

- Listening - to gain maximum benefit from a question, we must listen to the response. This may require us to remove distractions and show some empathy for the interviewee. Don't forget that if we are to listen, we have to stop talking – this may be easier for some than others!

- Observations – be prepared to stop and watch, and if you do not understand what is happening, ask. Always be prepared to follow up a suitable response with a request to 'show me' the evidence. Always try and see, and record, objective evidence.

Exit meeting

Exit meetings must be held. Even if you did not give notice, everyone by now knows what you are doing and is interested in your findings and recommendations. Before you leave, meet with whoever is in charge and give feedback. It is always important to discuss the non-conformances and agree what can be done to correct them in terms of action and timescale.

If you cannot agree at the exit meeting, record that on the inspection checklist.

c) Reporting

The report will need to meet organisational needs, but in the interests of brevity there is usually little need to reproduce your checklist. However, the complete checklist should be retained as it is dated evidence of what was good or bad.

The introduction to the report should detail the premises, workplace and those present at the inspection.

The body of the report is a summary of the findings, giving information on the findings, both good and bad. Do compliment the things that you have seen that were good, and include everything that needs improvement. For those items, detail the agreed action and timescale.

If the report is of a type that includes scoring, you may wish to indicate whether or not there has been an improvement.

d) Follow up

This allows the inspector to confirm that corrective action has been completed. How it will take place may have been agreed at the exit meeting. It may be a written confirmation, a copy of an order or delivery note etc., but may just be a visit to see the new process etc. Usually, but depending on risk, this will wait until the next inspection.

If corrective action has not taken place by the agreed time, the reason must be reported to the relevant risk management group or manager and included on the next inspection.

The process of inspecting premises is a vital part of effective management. If the process as detailed above is used, action planning and true accountability can be brought to any organisation, large or small.

14 Consultation with Employees

When The Health and Safety at Work Act was being prepared it was recognised that it was vital to involve everyone in the workplace.

Having in place an understanding, and a process to keep everybody informed and consulted on health and safety matters, is often known as the 'health and safety culture of the organisation.' Supervisors and managers have an important role in making this culture a reality.

CONSULTATION

Consultation has been defined as 'the active consideration of others' views'. Effective consultation requires a communication process that involves everyone and acts on the messages that result.

In small organisations this is relatively easy, given the commitment of managers. Health and safety becomes a topic of conversation at staff meetings, on purchasing decisions and when considering job design.

In large organisations it is more difficult as it depends on the enthusiasm and commitment of local managers and supervisors to involve all employees. Methods that have proved to be successful include:

- *Team brief systems,* where organisation-wide information is combined with local matters and given at staff meetings, allowing feedback.
- *Health and safety newsletters,* highlighting good practice and providing a route for suggestions, are often used. These can pass on information about changes to legislation or practice, advertise training sessions, etc. They should be interesting to read, perhaps with competitions that have prizes.
- *Pay packet notices* ensure that everyone receives the relevant information, and if an opportunity is given for suggestions, they can be very effective.
- *Local consultation* is the best method, where supervisors meet with their colleagues to discuss health and safety matters. Safety representatives are a valuable resource at these meetings and if they co-operate and work with the supervisor, everyone's views can be taken into account when decisions are made.

LEGAL REQUIREMENTS

Although legislation exists to require consultation, it is in every organisation's interest to promote the health and safety culture.

The Health and Safety at Work Act etc. 1974 requires employers to consult with safety representatives to enable employers and employees to co-operate effectively in promoting and developing measures to ensure the health and safety of the employees.

Safety representatives are required to represent the employees in these consultations.

The Management of Health and Safety at Work Regulations 1999 reinforce this requirement by prescribing occasions when the employer must consult.

The Safety Representatives and Safety Committee Regulations 1977 gave a framework in which consultation could be managed in workplaces where Trades Unions are active. **The Health and Safety at Work (Consultation with Employees) Regulations 1996** provide similar rules for employers who have no recognised trades unions.

THE SAFETY REPRESENTATIVES AND SAFETY COMMITTEE REGULATIONS 1977

Safety representatives

A recognised trade union may appoint safety representatives from amongst the employees - the safety representative has usually either been employed for two years or has two years' experience in similar employment.

By agreement, a safety representative may act on behalf of members of other unions. Numbers of representatives are agreed using existing industrial relations procedures with recognised trades unions.

Functions of a safety representatives

a) to investigate potential hazards and dangerous occurrences at the workplace, and to examine causes of accidents at the workplace;

b) to investigate complaints by any employee he represents relating to that employee's health, safety or welfare at work;

c) to make representations to the employer on matters arising from a) & b) above;

d) to make representations to the employer on general matters affecting the health, safety or welfare at work of the employees at the workplace;

e) to carry out inspections;

f) to represent employees in consultations with the enforcers of the health and safety legislation;

g) to receive information from inspectors; and

h) to attend safety committee meetings.

Time off with pay shall be provided for these functions and for reasonable training to carry out those responsibilities.

Inspections

Safety representatives are entitled to inspect the workplace on a three-monthly basis after giving reasonable notice in writing (which may be an agreed programme).

More frequent inspections may be carried out with the agreement of the employer and following substantial changes to conditions of work or publication of relevant new guidance etc. from HSE. Representatives are entitled to reasonable co-operation from the employer during inspections.

Where safety representatives have made a written report to the employer, the employer will normally take appropriate action. If not, or the representative is not satisfied with the action, the employer should give an explanation in writing to the representative. This should be publicised throughout the workplace.

Where there has been a notifiable accident or dangerous occurrence, or a notifiable disease contracted, representatives may carry out an inspection of the workplace concerned. The main purpose of this inspection shall be to determine the cause to prevent recurrence.

Representatives are entitled to see and take copies of any record required under the health and safety legislation, except if:

a) it is identifiable to an individual (unless consent given) or

b) the information has been obtained to bring, prosecute or defend legal proceedings. This includes all risk assessments, accident reports etc.

Safety committees

If two representatives request in writing, the employer should establish a safety committee within three months. Consultation with the representatives should take place on the way in which the committee will function. In large workplaces it may be necessary to set up several committees with adequate arrangements for co-ordination between them.

Safety committees have the function of keeping under review the measures taken to ensure the health and safety at work of employees. Agreed objectives and terms of reference should be drawn up to facilitate that function.

Other functions may be defined. These might include:

a) the study of accident and disease statistics and trends;

b) examination of safety reports;

c) consideration of inspectors' reports;

d) consideration of safety representatives' reports;

e) assistance in the development of safe systems of work;

f) monitoring effectiveness of health and safety training;

g) monitoring effectiveness of health and safety communication and publicity in the workplace; and

h) providing a link with the enforcing authority.

The safety committee functions must not be confused with the responsibility of management to take executive action - the role of the committee is to supplement the management action.

The size of the committee should be as compact as possible, but with adequate representation of the interests of management and all employees. The number of management representatives should not exceed the number of employees' representatives.

Safety committees should meet as often as necessary, depending on volume of business. Sufficient time should be allowed to ensure full discussion of all business. Meeting dates should be arranged well in advance and all members of the committee given a copy of the programme. Dates should be published for all employees to see, and agenda and supporting papers sent to all members at least one week before each meeting.

Minutes should be taken, agreed and a personal copy supplied to each member as soon as possible after each meeting. A copy should be sent to each safety representative and to the most senior executive.

THE HEALTH AND SAFETY (CONSULTATION WITH EMPLOYEES) REGULATIONS 1996

These regulations require consultation with employees not covered by trade union safety representatives. This may be either directly with employees or by 'representatives of employee safety' (i.e. elected by the employees they represent).

The requirements top up the consultation process through trade unions – for example, if a trade union safety representative has not been appointed the regulations also apply.

Similar functions and time off with pay apply to representatives of employee safety. However, training has to be provided by the employer, as there is no trade union to support the representative.

15 Personal Protective Equipment

Personal protective equipment (PPE) is designed to protect the wearer from one or more risks. It includes aprons, protective clothing for poor weather, gloves, safety footwear, safety helmets, high visibility clothing, eye protectors, respirators and safety harnesses etc. The function is to reduce risk for the wearer. PPE should only be used after a risk assessment has been carried out and it has been decided that it is the only way of controlling risk.

Most uniforms, clothing for the food industry, protective clothing for competitive sports and motor cycle helmets are not included in the regulations.

PERSONAL PROTECTIVE EQUIPMENT AT WORK REGULATIONS 1992

Regulation 4 **Every employer shall ensure that suitable personal protective equipment is provided to his employees who may be exposed to a health and safety risk at work.**

In the hierarchy of control measures for risks to health and safety, personal protective equipment (PPE) is always regarded as the last resort. Engineering controls or safe systems of work should always be considered first.

We have to remember that PPE protects only the person wearing or using it, whereas controlling risks at source protects everyone. PPE may also restrict the movement of the wearer, hinder the job being done or be uncomfortable to wear.

The provision of PPE means that it must be readily available and that there is no charge on the employee for it. An example of this is safety footwear - if there is a risk of danger to feet by the dropping of objects, then footwear should be provided free of charge. Other employees may however wish to wear safety shoes at their own expense.

PPE displaying the 'CE' mark must have passed basic performance tests. It is an offence to sell PPE that does not carry the 'CE' mark.

Regulations 6, 7 & 8

The employer must ensure that PPE is suitable, maintained and stored correctly.

This regulation is to ensure that PPE is fit for the purpose. In some cases the assessment does not require recording; in more complex cases it does need to be recorded and be accessible to those who need it.

PPE should be subject to a system of maintenance and checking to ensure that it maintains protection. This could range from a user's weekly visual check to an accurate record of wearing times and competent person checks. Storage conditions should protect the equipment from contamination, loss or damage.

Regulations 9 & 10

If PPE is provided the employee must have information, instruction and training.

The employer must ensure that PPE is worn correctly.

Employees must use PPE as instructed.

Employers should ensure that employees are able to make effective use of PPE to protect them against hazards at work. The recommendations of the manufacturer should be taken into account and information and training provided in a manner that is easily understood by the user. Training should be theoretical and practical and be repeated at appropriate intervals.

Employers should check that PPE is being used correctly and employees have a duty to use it correctly.

TYPES OF PPE

Hand protection

There are many types of glove, from the latex or vinyl gloves used to protect against infection to chain mail gauntlets used in butchery.

Eye protection

These can protect against impact, splashes, mists and sprays, dust, welding arc and laser light. The choice of design is enormous, but the employer must ensure that the correct protection is offered. Eye protection is marked with series of letters and signs that give information on the level of protection from the eyewear.

Foot protection

Safety shoes can include toe and sole protection. These are available in many designs and level of comfort. Wellingtons and wet weather footwear, clogs, anti-static and anti-slip footwear is also included.

Respiratory protective equipment

It is important that if respiratory protective equipment (RPE) is used a full risk assessment has been carried out and a decision made that RPE is the only way of reducing risk to the worker.

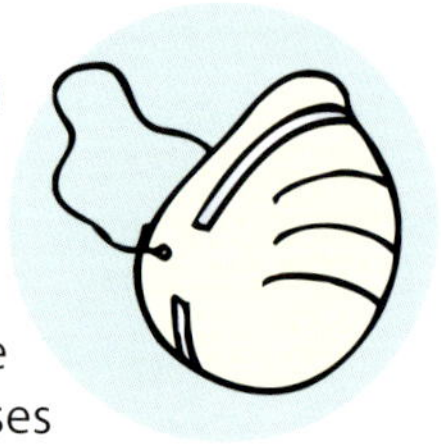

Face masks are very simple filters that remove relatively large particles of dust. They do not offer protection against fine particles, gases or fumes.

Respirators incorporate a filter designed to remove fine solids. These can be face mask type, with a tight fitting to the face, or can include a pumped supply of air through a filter to a mask or helmet.

Gas respirators incorporate a cartridge or canister that filters specific chemicals from air. Care must be taken to use the correct filter and ensure that it does not become saturated.

Breathing apparatus provides a supply of breathable air to the wearer. It can be used in dangerous atmospheres, such as confined spaces.

16 Safety Signs

HEALTH AND SAFETY (SAFETY SIGNS AND SIGNALS) REGULATIONS 1996

The regulations cover various means of communicating health and safety information. These include the use of illuminated signs, hand and acoustic signals (e.g. fire alarms), spoken communication and the marking of pipework containing dangerous substances. These are in addition to traditional signboards such as prohibition and warning signs. Fire safety signs (i.e. signs for fire exits and fire-fighting equipment) are also covered.

They require employers to provide specific safety signs wherever there is a risk that has not been avoided or controlled by other means, e.g. by engineering controls and safe systems of work. Where a safety sign would not help to reduce that risk, or where the risk is not significant, there is no need to provide a sign.

Safety signs are helpful, but we all have to be realistic and understand that after a few days, signs lose effectiveness. They seem to blend into the background of decorations and other notices.

Examples of good signage in use.

Green Signs indicate a SAFE condition.

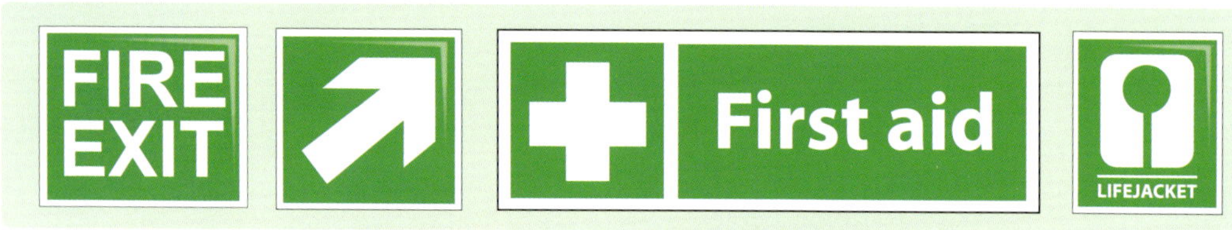

Red Signs indicate a PROHIBITION.

Blue Signs indicate MANDATORY instruction.

Yellow Signs indicate a HAZARD or danger.

17 Examinations

THE EXAMINATION

There are various examining bodies for health and safety in the UK, and they may use written examination scripts or multiple choice questions. The CIEH Intermediate Certificate in Supervising Health and Safety examination paper consists of 60 questions and you are required to answer them all. Each question is worth one mark.

Standard examination conditions mean that you will not be allowed to talk to other candidates, refer to any text books or notes and you must work in silence. If you should require assistance of any kind during the examination you must ask the invigilator. These conditions apply to give you the best chance to succeed.

The best piece of examination advice you could ever be given is to read all the parts of each question carefully before you choose your answer.

Answer Information Sheets are often scanned and marked electronically so it is essential that you complete both the general information and the answer information legibly and in the correct manner using a 2B pencil and erasing any mistakes thoroughly.

You will have time at the end of the examination to review your responses. For multiple choice papers it is essential that if you change your answer choice, the original answer must be thoroughly erased and the new answer choice clearly marked by crossing the box next to the chosen option.

MULTIPLE CHOICE EXAMINATION QUESTIONS

The questions are not designed to trip you up and there are no trick questions. There is always just one correct answer. There are four answer options and you are required to choose only one. If you mark more than one answer on your answer sheet and it is not absolutely clear which is your final choice then no marks will be awarded for that question.

Important words in the question are in upper case type, e.g. 'Give one HUMAN factor...' Read the question and answer options carefully before marking your answer on the answer sheet. Each question is worth one mark.

The most common reasons for losing marks are:

- Not reading the question properly. You should make sure you understand the question before you choose your answer. Try to identify the important words in the question and think carefully before you mark your choice on the answer sheet. Read the whole question first before choosing an answer as you are less likely to misunderstand the question and make the wrong choice.
- If you have marked more than one answer option no marks will be awarded for that question.
- Never leave a question unanswered. If you are not sure of the answer, as a last resort you should make an educated guess.
- Some candidates do not manage their time well and find they cannot complete the examination in the allocated time. Keep a regular check on the clock and if you are having difficulty answering a question move on to the next question and then revisit the question later.

Glossary

ABSOLUTE DUTIES Certain duties under health and safety law are of an absolute or strict nature. These duties are qualified by the terms 'shall' or 'must'. For instance, under The Workplace (Health, Safety and Welfare) Regulations 1992 a workplace and the equipment, devices and systems shall be maintained in an efficient state, in efficient working order and in good repair.

ACCIDENT The term 'accident' can be defined in a number of ways:

- an unforeseeable event, often resulting in injury;
- an unplanned and uncontrolled event, which has led to or could have caused injury to persons, damage to plant or other loss; and
- a deviation from the normal, the expected or the planned, usually resulting in injury.

ACCIDENT STATISTICS Accident statistics, as a reactive form of safety monitoring, generally take the form of an incident rate.

ACTIVE MONITORING This form of monitoring entails the carrying out of a range of proactive safety monitoring exercises directed at preventing accidents, such as safety audits and safety inspections.

ACT OF PARLIAMENT (STATUTE) This is the principal form of legislation that has gone through the parliamentary process to become statute, e.g. The Health and Safety at Work etc. Act 1974 (HSWA).
The HSWA enables the Secretary of State for Employment to make regulations on a range of health and safety related matters, e.g. The Noise at Work Regulations 1989.

ADJUSTABLE GUARD A guard incorporating an adjustable element, which, once adjusted, remains in that position during a particular operation.

ALTERNATIVE MEANS OF ESCAPE This is a second route, usually in the opposite direction to the first means of escape in the case of fire, which may join the first means of escape.

APPROVED CODE OF PRACTICE (ACOP) The Health and Safety Commission (HSC) is empowered under the HSWA to approve and issue codes of practice for the purpose of supplementing information contained in statute and regulations. A code of practice can be drawn up by the HSC or some other body. In every case, however, any relevant government department or other body must be consulted beforehand and the approval of the Secretary of State for Employment must be obtained.
Failure to comply with an ACOP can raise the presumption that the offender was in breach of the legislation unless 'works of an equivalent nature' are undertaken, which satisfy the requirements of the ACOP.

ARM PROTECTION This may take the form of gauntlets, chain mail and leather arm protectors. This form of protection is commonly used where operators are using knives or other implements where there is a risk of arm injury.

ARTICLE FOR USE AT WORK
This means:
- any plant designed for use or operation (whether exclusively or not) by persons at work, or persons who erect or install any fairground equipment; and
- any article designed for use as a component in any such plant or equipment (Health and Safety at Work etc. Act 1974).

ATTITUDE
Variously defined as:
- a predetermined set of responses built up as a result of experience of situations; and
- a tendency to respond favourably or unfavourably to certain persons, objects or situations.

AUTOMATIC GUARD
A guard that is associated with, and dependent upon, the mechanism of the machinery and operates so as to remove physically from the danger area any part of a person exposed to the danger.

BODY PROTECTION
This form of protection may include one-piece and two-piece overalls, donkey jackets, aprons, warehouse coats, foul weather clothing, and clothing to protect against machinery, together with high-visibility clothing, life-jackets and buoyancy aids.
Certain forms of body protection can be of the washable, semi-disposable or disposable type, such as overalls.

BREACH OF STATUTORY DUTY
In certain situations a breach of a criminal duty may give rise to civil liability.

BREATHING APPARATUS
This is generally taken to mean apparatus that replaces the surrounding air supply and provides the user with sufficient air to breathe normally. The use of breathing apparatus may be necessary in atmospheres contaminated with smoke and other airborne pollutants.

CARBON DIOXIDE EXTINGUISHER
This extinguisher consists of a pressure cylinder filled with liquid carbon dioxide. On operation of the trigger the liquid is discharged through the horn under its own pressure, being converted to carbon dioxide snow prior to application.

CHEMICAL FOAM EXTINGUISHER
These appliances incorporate a cylinder containing two solutions that are mixed on activation. The foaming mixture is expelled by carbon dioxide generated in the chemical reaction.

CIVIL LIABILITY
This refers to the 'penalty' that can be imposed by a civil court, e.g. County Court, High Court.
A civil action generally involves individuals, with a plaintiff suing a defendant for a remedy or remedies. In most cases, the remedy takes the form of damages, a form of financial compensation.
Civil actions commonly include allegations of negligence and/or breach of statutory duty.

COMFORT VENTILATION This is the process of providing sufficient air for people to breathe and, to some extent, regulating temperature. It is directly related to the number of air changes per hour in a workplace according to the external ambient air temperature and the actual rate of air movement. Rates of air change will, in most cases, vary from summer to winter to maintain comfort ventilation.

COMPETENT PERSONS The appointment by an employer of suitable competent persons to undertake a range of examinations, inspections, operations and supervisory duties is well-recognised in health and safety law. Generally, a competent person should have the appropriate skill, knowledge and experience to deal with the particular circumstances and be appointed by the employer.

Under the MHSWR 1999, an employer must:

- appoint one or more competent persons to assist him in undertaking the measures he needs to take to comply with the requirements and prohibitions imposed by or under the relevant statutory provisions; and
- appoint competent persons to oversee the emergency evacuation procedures.

'CORROSIVE' CLASSIFICATION Substances and preparations that may, on contact with living tissue, destroy them.

CRIMINAL LIABILITY A crime is an offence against the state.

Criminal liability refers to the responsibilities of persons under statutes and regulations and the penalties that can be imposed by the Criminal Courts, e.g. Magistrates' Court and Crown Court.

The burden of proving a criminal charge 'beyond reasonable doubt' rests with the prosecution.

Where a person is found guilty, a court will impose some form of punishment, namely, a fine and/or imprisonment.

DANGER Liability or exposure to harm; a thing that causes peril.

DANGEROUS OCCURRENCE An event listed in Schedule 2 of The Reporting of Injuries, Diseases and Dangerous Occurrences Regulations 1995 (RIDDOR). Such events must be notified immediately to the enforcing authority and reported on Form F2508. Examples of dangerous occurrences are the collapse or overturning of lifting machinery, unintentional explosions, gassing accidents and boiler explosions.

DEADSTOP DEVICE A form of safety trip device**,** incorporating a trip bar or arm, an electrical switch and brake, located in close proximity to a danger point on a machine. When the trip arm, e.g. a telescopic arm to a vertical drill, is touched, the machine motor automatically cuts out.

DISABILITY GLARE The visually disabling effect produced by bright bare lamps directly in the line of sight (dazzle).

DISCOMFORT GLARE A form of glare caused by too much contrast of brightness between an object and its background.

DISPLAY SCREEN EQUIPMENT Any alphanumeric or graphic display screen, regardless of the display process involved.

(The Health and Safety (Display Screen Equipment) Regulations 1992.)

DISTANCE GUARD A guard that does not completely enclose a danger point or area but that places it out of normal reach.

DRY POWDER CHEMICAL EXTINGUISHER These appliances are of two types, stored pressure and gas cartridge. In both cases, the cylinder is fed with dry powder.
In the stored pressure type, the cylinder is pressurised with dry air or nitrogen, and the operation of the trigger allows the pressure to expel the dry powder through the hose.
With the gas cartridge type, the seal on the cartridge is broken by a plunger and the flow of powder is controlled by a trigger valve on the hose.

EAR PROTECTION Ear or hearing protection is designed to protect people from the risk of noise-induced hearing loss (occupational deafness). It takes three principal forms:

- ear plugs: fitted into the auditory canal;
- ear defenders, muffs and pads: cover the whole ear and can reduce exposure by up to 50 dBA at certain frequencies; and
- ear valves: inserted into the auditory canal, allowing normal conversation to take place while preventing harmful noise reaching the ear.

EMERGENCY LIGHTING This is provided to ensure safety when a normal lighting installation fails, and may take the form of standby lighting or escape lighting.
Standby lighting enables essential work to continue, the level of illuminance depending upon the nature of the work. This may be between 5 and 100% of the illuminance provided by the normal installation,.
Escape lighting enables a building to be evacuated safely and may take the form of battery or generator-powered installations.

EMPLOYEES' DUTIES Under the HSWA 1974, employees must:

- take reasonable care for the safety of themselves and other persons who may be affected by their acts or omissions whilst at work;
- cooperate with their employer so far as is necessary for him to comply with any duty or requirement under any of the relevant statutory provisions; and
- not intentionally or recklessly interfere with or misuse anything provided for the purposes of health and safety at work in furtherance of a statutory requirement.

Under the MHSWR 1999, employees must:

- use any machinery, equipment, dangerous substance, transport equipment, means of production or safety device provided to him by his employer in accordance with any training received by him in the use of the equipment concerned and the instructions regarding that use that have been provided to him by his employer in compliance with the requirements and prohibitions imposed upon that employer by or under the relevant statutory provisions; and
- inform their employer or any other employee with specific responsibility for the health and safety of his fellow employees:
    - of any work situation that represents a serious and immediate danger to health and safety; and
    - of any matter that represents a shortcoming in the employer's protection arrangements for health and safety.

EMPLOYER'S DUTIES Under the HSWA 1974 it is the duty of every employer, so far as is reasonably practicable, to ensure the health, safety and welfare at work of all his employees. More particularly, this includes:

- the provision and maintenance of plant and systems of work that are, so far as is reasonably practicable, safe and without risks to health;
- arrangements for ensuring, so far as is reasonably practicable, safety and absence of health risks in connection with the use, handling, storage and transport of articles and substances;
- the provision of such information, instruction, training and supervision as is necessary to ensure, so far as is reasonably practicable, the health and safety of all employees;
- so far as is reasonably practicable as regards any place of work under the employer's control, the maintenance of it in a condition that is safe and without health risks, and the provision and maintenance of means of access to and egress from it that are safe and without health risks;
- the provision and maintenance of a working environment for their employees that is, so far as is reasonably practicable, safe, without risk to health and adequate as regards facilities and arrangements for their welfare at work;
- employers must prepare and, as often as is necessary, revise a written Statement of Health and Safety Policy, and bring the Statement and any revision of it to the notice of all their employees; and
- every employer must consult appointed safety representatives with a view to making and maintaining arrangements that will enable them and their employees to co-operate effectively in promoting and developing measures to ensure the health and safety at work of the employees, and in checking the effectiveness of such measures.

Under the MHSWR 1999 every employer must:

- make a suitable and sufficient assessment of the risks to his own employees and other persons affected by his activities in order to identify the measures he needs to take to comply with the requirements and prohibitions imposed on him by or under the relevant statutory provisions;
- review and revise risk assessments and implement changes where necessary;
- ensure the effective planning, organisation, control, monitoring and review of the preventive and protective measures arising from the risk assessment;
- provide health surveillance where necessary;
- appoint one or more competent persons to assist him in complying with his duties under the relevant statutory provisions;
- establish and, where necessary, give effect to procedures to be followed in the event of serious or imminent danger, and nominate competent persons to implement these procedures;
- provide employers with comprehensible and relevant information on specific aspects;
- where a workplace is shared, co-operate with other employers, take all reasonable steps to co-ordinate safety procedures, and inform other employers of the risks arising out of or in connection with his own undertaking;
- provide comprehensible information to employees from an outside undertaking on the risks and measures he has taken to comply with the relevant statutory provisions;
- take into account health and safety capabilities when allocating tasks;

- ensure health and safety training of employees:
  - on recruitment; and
  - on being exposed to new or increased risks due to transfer, change of responsibility, introduction of new work equipment, change in existing work equipment, introduction of new technology, introduction of a new system of work or change of an existing system;
- provide temporary workers with certain comprehensible information; and
- consult with safety representatives in specified circumstances.

EMPLOYER'S LIABILITY

Under the Employer's Liability (Compulsory Insurance) Act 1969 employers are required to carry insurance against common law liability to their employees resulting in injury, disease and/or death. Failure to do so is a criminal offence. A current employer's liability insurance certificate must be displayed in a prominent position at the workplace.
The policy must state that any person under a contract of service or apprenticeship who sustains injury, disease or death caused during the period of insurance and arising out of the course of employment will be covered for any legal liability on the part of the employer to pay compensation.

ENFORCING AUTHORITY

The enforcing authorities are:

- the Health and Safety Executive (HSE);
- local authorities; and
- fire authorities for certain delegated matters.

Enforcement of health and safety legislation is undertaken by inspectors appointed under section 20 of the HSWA and authorised by written warrant from the enforcing authority.

ESCAPE LIGHTING

(see EMERGENCY LIGHTING.)

EVACUATION

The process of moving people from a building or area to an identified assembly point.

'EXPLOSIVE' CLASSIFICATION

Solid, liquid, pasty or gelatinous substances and preparations that may react exothermically without atmospheric oxygen, thereby quickly evolving gases and that, under defined test conditions, detonate, quickly deflagrate or, upon heating, quickly explode when partially confined.

'EXTREMELY FLAMMABLE' CLASSIFICATION

Liquid substances and preparations having an extremely low flashpoint and a low boiling point and gaseous substances and preparations that are flammable in contact with air at ambient temperature and pressure.

EYE PROTECTION

Eye protection is designed to prevent injury arising from flying particles, dust, fumes, welding glare and splashes of hazardous substances, such as acids.
Eye protection may take a number of forms:

- safety spectacles: with toughened glass or plastic lenses with plastic or metal frames and, in some cases, incorporating side shields;
- safety goggles: which incorporate a cup surrounding each eye; and
- eye and face shields: which can be hand-held, fixed to a helmet or strapped to the head.

FACE MASKS

These generally incorporate a metal frame for holding a filtering medium against the nose and mouth to protect against inhalation of coarse dust particles or non-toxic paint sprays only. They afford no protection against gases, fumes, mists and other forms of airborne contamination.

FACE SHIELDS These provide protection to the eyes and face and can be hand-held, fixed to a helmet or strapped to the head. They are manufactured in toughened polycarbonate materials.

FIRE 'Fire' can be defined in several ways:

- a spectacular example of a fast chemical reaction between a combustible substance and oxygen accompanied by the evolution of heat;
- a mixture in gaseous form of a combustible substance and oxygen with sufficient energy put into the mixture to start a fire; and
- an unexpected combustion generating sufficient heat or smoke resulting in loss of life, damage to plant, equipment, goods and buildings.

FIRE ALARM The most effective fire alarm is the human voice. Fire alarm systems can range from a simple hand bell to electronically actuated systems. Systems are commonly linked with heat and/or smoke detection systems, the detection system actuating the audible alarm and sprinkler systems.

FIRE CERTIFICATE Under The Fire Precautions Act 1971 certain classes of industrial and commercial premises must hold a fire certificate.
A fire certificate specifies:

- the use or uses of the premises it covers;
- the means of escape in case of fire indicated on a plan of the premises;
- the means for ensuring the safety and effectiveness of the means of escape, such as fire and smoke stop doors, emergency lighting and direction signs;
- the means of fighting fire for the use of persons on the premises;
- the means of raising the alarm; and
- in the case of factories, particulars of explosive or highly flammable substances stored and used on the premises.

A fire certificate may also impose requirements relating to:

- the maintenance of the means of escape and for keeping it free from obstruction;
- the maintenance of other fire precautions;
- the training of people and the keeping of records;
- limitations on numbers of persons in the premises at any time; and
- any other relevant fire precautions.

FIRE DRILL A planned or unplanned evacuation of a building or area in order to train and familiarise occupants with the means of escape.

FIRE ESCAPE A commonly used term to indicate the structural means of escape from a building, such as an external metal staircase or fire-protected internal staircase.

FIRE EXIT A designated door giving direct access to the open air and forming part of a means of escape in the event of fire.

FIRE EXTINGUISHERS

These are appliances designed to be carried and operated by hand. They contain an extinguishing medium that can be expelled by the action of internal pressure and directed at a fire.
The pressure may be stored or obtained through a chemical reaction or by the release of gas from a cartridge.
All portable extinguishers must be coloured red and display a distinguishing label thus:

Water	Red
Foam	Cream
Carbon dioxide	Black
Dry chemical powder	Blue
Vaporising liquid	Green

FIRE INSTRUCTIONS

A fire instruction is a notice informing people of the action they should take on either hearing a fire alarm or discovering a fire.

FIRE TRIANGLE

There are three requirements for the propagation of a fire, commonly known as the 'fire triangle', namely:

- Oxygen or air;
- A fuel or combustible substance; and
- A source of energy, heat or ignition.

The removal of one of the three elements of the fire triangle results in combustion failing to take place.

FIRST AID

The skilled application of accepted principles of treatment on the occurrence of an accident or in the case of sudden illness, using facilities and materials available at the time.
First aid is defined by law as follows.

- in cases where a person will need help from a medical practitioner or nurse, treatment for the purpose of preserving life and minimising the consequences of injury or illness until such help is obtained; and
- treatment of minor injuries that would otherwise receive no treatment or that do not need treatment by a medical practitioner or nurse.

(The Health and Safety (First Aid) Regulations 1981.)

FIXED GUARD

A guard that has no moving parts associated with it, or dependent upon the mechanism of any machinery and that, when in position prevents access to a danger point or area.

FOAM EXTINGUISHERS

These extinguishers take the form of chemical foam and stored pressure foam extinguishers.

FOOT PROTECTION

Foot protection includes safety boots, shoes, clogs, foundry boots, wellington boots, anti-static footwear and ankle protection. Foot protection should be waterproof, resistant to acids, alkalis, oils and other substances. It should incorporate steel toecaps where there is a risk of foot injury. With wet processes, wellington boots, incorporating a steel toecap, should be provided.

GENERIC (MODEL) RISK ASSESSMENT
A risk assessment produced once only for an identified type of activity, workplace or work group. For generic risk assessments to be effective:
- 'worst case' situations must be considered; and
- provision must be made in the assessment document to monitor the implementation of the controls that are or are not relevant in the case of a particular activity, workplace or work group.

GLARE
The effect of light that causes discomfort or impaired vision, which is experienced when parts of the visual field are excessively bright compared with the general surroundings.

GUARD
A physical barrier to protect persons from danger.
(**See FIXED GUARD, ADJUSTABLE GUARD, DISTANCE GUARD, INTERLOCKING GUARD and AUTOMATIC GUARD.**)

HAND PROTECTION
This includes gloves and gauntlets, principally concerned with protecting the hands from injuries such as cuts, abrasions and burns and those arising during manual handling operations and in extremes of temperature.

'HARMFUL' CLASSIFICATION
Substances and preparations that may cause death or acute or chronic damage to health when inhaled, swallowed or absorbed via the skin.

HAZARD
Something with the potential to cause harm. This can include substances or machines, methods of work and other aspects of work organisation.

HEAD PROTECTION
This form of protection includes:
- safety helmets to protect the head from falling objects and overhead hazards;
- industrial scalp protectors (bump caps) to protect against striking fixed obstacles, scalping or entanglement; and
- caps and hairnets to prevent the hair from coming into contact with moving machinery.

HEALTH AND SAFETY AT WORK etc. ACT 1974
The HSWA covers all people at work except domestic workers in private employment and extends to the prevention of risks to the health and safety of the general public. The general objectives of the HSWA are:
- to secure the health, safety and welfare of all persons at work;
- to protect others from the risks arising from workplace activities;
- to control the obtaining, keeping and use of explosive or highly flammable substances; and
- to control emissions into the atmosphere of noxious or offensive substances.

All duties are qualified by the term 'so far as is reasonably practicable.'
(**See EMPLOYEES' DUTIES, EMPLOYER'S DUTIES, REASONABLY PRACTICABLE.**)

HEALTH AND SAFETY EXECUTIVE
This is the national health and safety enforcing authority. HSE inspectors have wide-ranging powers, including the power to serve Improvement Notices and Prohibition Notices.
(See IMPROVEMENT NOTICE, POWER OF INSPECTORS, PROHIBITION NOTICE.)

HERNIA

A protrusion of an organ from one compartment of the body into another, e.g. a loop of intestine into the groin or through the frontal abdominal wall.
Both these forms of hernia can result from incorrect manual handling techniques and particularly from the adoption of bent back stances, which produce compression of the abdomen and lower intestines.

'HIGHLY FLAMMABLE' CLASSIFICATION

The following substances and preparations, namely:

- substances and preparations that may become hot and finally catch fire in contact with ambient temperature without any application of energy;
- solid substances and preparations that may readily catch fire after brief contact with a source of ignition and that continue to burn or to be consumed after removal of the source of ignition;
- liquid substances and preparations that have a very low flashpoint; and
- substances and preparations that, in contact with water or damp air, evolve highly flammable gases in dangerous quantities.

HOISTS AND LIFTS

General requirements for hoists and lifts are incorporated in The Lifting Operations and Lifting Equipment Regulations 1998.
Every hoist or lift shall be:

- of good mechanical construction, sound material, adequate strength and properly maintained;
- examined by a competent person every six months at least, recorded and, where defects are found, a report must be sent to the HSE;
- every hoistway or liftway shall be efficiently protected by a substantial enclosure fitted with gates of such design and layout as to prevent, when the gates are shut, any person falling down the way or coming into contact with any moving part of the equipment. Any such gate must be fitted with an efficient interlocking device;
- every hoist or lift and every enclosure shall be so constructed as to prevent any part of any person or any goods carried from being trapped between any part of the hoist or lift and any fixed structure or between the counterbalance weight or any other moving part; and
- the safe working load of the hoist or lift must be conspicuously marked and must not be exceeded.

HSE GUIDANCE NOTES

The HSE issues Guidance Notes that have no legal status and are purely of an advisory nature, sometimes to supplement information in an ACOP.
HSE Guidance Notes are issued in six specific series:

- General Safety (GS);
- Chemical Safety (CS);
- Environmental Hygiene (EH);
- Medical Series (MS);
- Plant and Machinery (PM); and
- Health and Safety Guidance (HS(G)).

HUMAN ERROR

Limitations in human capacity to perceive, attend to, remember, process and act on information are all relevant in the context of human error.
Typically, human error is associated with:

- lack of knowledge;
- lack of concentration;
- incapability; and
- wilfullness.

HUMAN FACTORS
This term is used to cover a range of issues, including:
- the perceptual, physical and mental capabilities of people and the interaction of individuals with their job and working environments;
- the influence of equipment and system design on human performance; and
- the organisational characteristics that influence safety-related behaviour.

These issues are affected by:
- the system for communication within the organisation; and
- the training systems and procedures in operation, all of which are directed at preventing human error.

IGNITION
Ignition of a substance is brought about by the introduction of energy sufficient to raise a volume of the substance to its ignition temperature.
Ignition sources include open flames, electrical equipment, friction, matches, lighters, static electricity, lightning, welding and other hot work, engines and spontaneous ignition of certain flammable substances.

IMPROVEMENT NOTICE
An Improvement Notice is served on an employer and other persons when, in the opinion of an enforcement officer, a person, business or undertaking is not complying with the relevant statutory provisions and where action is required by a certain date.
There is a right of appeal to an industrial tribunal against the notice, which suspends its enforcement until a decision is reached by the tribunal.

INCIDENT
An event that does not result in injury, damage or loss but that may cause interruption of the work process.

INGESTION
The taking into the body of substances through the mouth.

INHALATION
The principal route of entry of hazardous substances, such as gases, dusts and vapours, into the body via the nose, mouth and respiratory tract.

INJECTION
A forceful breach of the skin, perhaps as a result of injury, which can transmit hazardous substances past the skin barrier.

INJURY
Damage to the body, commonly as a result of an accident.

INTERLOCKING GUARD
A guard that has a movable part so connected with the machinery controls that:
- the part(s) of the machinery causing danger cannot be set in motion until the guard is closed;
- the power is switched off and the motion braked before the guard can be opened sufficiently to allow access to the dangerous parts; and
- access to the danger point or area is denied while the danger exists.

'IRRITANT' CLASSIFICATION
Non-corrosive substances and preparations that through immediate, prolonged or repeated contact with the skin or mucous membrane may cause inflammation.

JOB SAFETY ANALYSIS
A technique that identifies all the accident prevention measures appropriate to a particular job or area of work activity, and the behavioural factors that most significantly influence whether or not these measures are taken.

LOW VOLTAGE — This is a protective measure against electric shock; the most commonly used reduced low voltage system is the 110 volt centre point earthed system. With this system the secondary winding of the transformer providing the 110 volt supply is centre tapped to earth, thus ensuring that at no part of the 110 volt circuit can the voltage to earth exceed 55 volts.

MACHINE — An apparatus for applying power, having fixed and moving parts, each with definite functions.

MANAGEMENT — The effective use of resources in the pursuit of organisational goals. 'Effective' implies achieving a balance between the risk of being in business and the cost of eliminating or reducing those risks.

MANDATORY SIGNS — Signs that indicate that a specific course of action is required, e.g. EYE PROTECTION MUST BE WORN. These signs are blue in colour, circular in shape, with a symbol or text in white.

MANUAL HANDLING OPERATIONS — This means any transporting or supporting of a load (including the lifting, putting down, pushing, pulling, carrying or moving thereof) by hand or by bodily force. (The Manual Handling Operations Regulations 1992.)

MEANS OF ESCAPE IN CASE OF FIRE — A continuous route by way of a space, room, corridor, staircase, doorway or other means of passage, along or through which persons can travel from wherever they are in a building to the safety of the open air at ground level by their own unaided efforts.

NEAR MISS — An unplanned and unforeseeable event that could have resulted, but did not result, in human injury, property damage or other form of loss.

NEGLIGENCE — 'Negligence' can be defined as 'careless conduct injuring another'. It has been defined at common law as:

- the existence of a duty of care owed by the defendant to the plaintiff;
- breach of that duty; and
- injury, damage or loss resulting or caused by that breach.

All the above circumstances must be established by an injured employee before a civil claim for damages can be considered by a court.

NOISE — Noise is generally defined as 'unwanted sound'.

NOISE-INDUCED HEARING LOSS (OCCUPATIONAL DEAFNESS) — Exposure to noise may affect hearing in three ways:

- temporary threshold shift: a short-term effect, i.e. a temporary reduction in the ability to hear, which may follow exposure to excessive noise, such as that from firework explosions;
- permanent threshold shift: a permanent effect where the limit of tolerance is exceeded in terms of the duration and level of exposure to noise and individual susceptibility; and
- acoustic trauma: a condition that involves sudden damage to the ear from short-term intense exposure or even from one single exposure, e.g. gunfire, major explosions.

'OXIDISING' CLASSIFICATION — Substances and preparations that give rise to an exothermic reaction in contact with other substances, particularly flammable substances.

PERMIT TO WORK A form of safe system of work operated where there is a high degree of foreseeable risk.
A formal safety control system designed to prevent accidental injury to personnel, damage to plant, premises and product particularly when work with a foreseeably high hazard content is undertaken and the precautions required are numerous and complex.

PERVASION The route of entry of hazardous substances through the skin.

PERSONAL PROTECTIVE EQUIPMENT All equipment (including clothing affording protection against the weather) that is intended to be worn or held by a person at work and which protects him against one or more risks to his health and safety. Any addition or accessory designed to meet this objective.
(The Personal Protective Equipment at Work Regulations 1992.)

POSTURAL FATIGUE A range of symptoms frequently experienced by users and operators of display screen equipment, in particular including lumbar, arm and shoulder pains, due to the adoption of unsuitable postures at a workstation. Generally, the more abnormal the posture, the greater the potential for postural fatigue and even long-term injury.

POWERS OF INSPECTORS Inspectors have the following powers:

- to enter premises at any reasonable time accompanied, if necessary, by a police officer;
- to take with them any duly authorised person, equipment or materials required;
- to make examinations and investigations;
- to direct that any premises, any part thereof or anything therein shall remain undisturbed for the purposes of examination and investigation;
- to take measurements, photographs, recordings and samples;
- to cause any article or substance to be dismantled or subjected to any process or test;
- to take possession of any article or substance and detain for as long as is necessary;
- to require any person to give information, answer questions and sign a declaration of truth;
- to require production of, inspect and take copies of books and documents required to be maintained or otherwise;
- to require any person to afford appropriate facilities and assistance;
- to inform safety representatives of matters they have found following an investigation or examination;
- to serve Improvement Notices and Prohibition Notices; and
- to prosecute offenders.

PRACTICABLE DUTIES A level of duty qualified by the term 'so far as is practicable' implies that the duty holder must take all measures necessary 'in the light of current knowledge or invention' or 'in the light of the current state of the art' to comply with the duty. 'So far as is practicable' implies a higher level of duty than 'so far as is reasonably practicable'. Cost or sacrifice cannot be considered in this case.

PROHIBITION NOTICE
Served by enforcement officers where they are of the opinion that a work activity involves or will involve a risk of serious personal injury.
The notice places a prohibition on specified activities until remedial measures are implemented.
It is not necessary for a legal requirement to have been contravened, but that there is an immediate threat to life.
A Prohibition Notice may be served with immediate effect or suspended for a period of time.
An appeal to an industrial tribunal does not suspend the implementation of the requirements outlined in the notice.

PROSECUTION
The bringing of a person before a court to answer a charge involving an alleged breach of the law. It is normal for the person charged to be served with a summons to attend court to answer to the charge or charges in question.

REACTIVE MONITORING
This includes a range of techniques, such as the investigation of accidents, ill-health and near misses, emergency and disaster planning, interpretation of statistical information and the implementation of a range of strategies to prevent recurrences of accidents and incidents.

REASONABLY PRACTICABLE
A level of duty qualified by the term 'so far as is reasonably practicable' implies that a computation must be made by the employer or owner in which the quantum of risk is placed on one scale and the sacrifice involved in the measures necessary for averting the risk (whether in money, time or trouble) is placed in the other, and that, if it be shown that there is a gross disproportion between them, the defendants discharge the onus on them.
(Edwards v National Coal Board (1949) 1 AER 743.)

REFLECTED GLARE
The reflection of bright light sources on shiny or wet work surfaces, such as glass or plated metal, which can almost entirely conceal the detail in or behind the object that is glinting.

RELEVANT STATUTORY PROVISIONS
The HSWA is an umbrella Act of general duties over, in some cases, former legislation, such as the Factories Act 1961 and the Offices, Shops and Railway Premises Act 1963. Both Acts and any regulations made under them are deemed to be the 'relevant statutory provisions' with regard to the HSWA. Schedule 1 of the HSWA lists regulations deemed to be the relevant statutory provisions. The Secretary of State for Employment may make regulations that are part of the relevant statutory provisions, e.g. the Control of Substances Hazardous to Health (COSHH) Regulations 2002.
Regulations implementing various European Directives in the UK form part of the relevant statutory provisions, e.g. the MHSWR 1999.

REPETITIVE STRAIN INJURY
This covers well-known conditions such as tennis elbow, flexor tenosynovitis and carpal tunnel syndrome, and is usually caused or aggravated by work associated with repetitive and over-forceful movement, excessive workloads, inadequate rest periods and sustained or constrained postures. This results in pain or soreness due to inflammation of muscles and the synovial lining of the tendon sheath.
Signs and symptoms include pain, tenderness, swelling and a grating sensation in the joint (crepitus) aggravated by pressure or movement.
(**See WORK-RELATED UPPER LIMB DISORDER.**)

REPORTABLE DISEASE

A disease or condition affecting a person at work and listed in Schedule 3 to The RIDDOR 1995. On diagnosis by a registered medical practitioner, such a disease or condition must be reported to the relevant enforcing authority by an employer on Form F2508A.

REPORTABLE INJURY

Under The RIDDOR 1995 where any of the following events arise out of work activities it must be notified by the quickest practicable means and subsequently reported within ten days on the appropriate form to the enforcing authority.
These events are:

- the death of any person as a result of an accident, whether or not they are at work;
- someone who is at work suffering a major injury as a result of an accident;
- someone who is not at work (e.g. a member of the public) suffering an injury as a result of an accident and is taken from the scene to a hospital for treatment, or if the accident happens at a hospital, suffering a major injury;
- someone at work is unable to do their normal work for more than three days as a result of an injury caused by an accident at work;
- the death of an employee, if this occurs some time after a reportable injury which led to the employee's death, but not more than one year afterwards; and
- a person at work suffering one of a number of specified diseases, provided that a doctor diagnoses the disease and the person's job involves a specified work activity.

A major injury is:

- any fracture, other than to the fingers, thumbs or toes;
- any amputation;
- dislocation of the shoulder, hip, knee or spine;
- loss of sight, whether temporary or permanent;
- a chemical or hot metal burn to the eye and any penetrating injury to the eye;
- any injury resulting from electric shock (including any electrical burn caused by arcing or arcing products) leading to unconsciousness or requiring resuscitation or admittance to hospital for more than 24 hours;
- any other injury:
  - leading to hypothermia, heat-induced illness or to unconsciousness;
  - requiring resuscitation; or
  - requiring admittance to hospital for more than 24 hours;
  - leading to loss of consciousness caused by asphyxia or by exposure to a harmful substance or biological agent;
- either of the following conditions that result from the absorption of any substance by inhalation, ingestion or through the skin:
  - acute illness requiring medical treatment; or
  - loss of consciousness; and
- acute illness that requires medical treatment where there is reason to believe that this resulted from exposure to a biological agent or its toxins or infected material.

RESPIRATORS

Respirators are designed to protect the respiratory system of the wearer against harmful dusts, gases, fumes and other forms of airborne contamination.

The correct selection of respiratory protection is vital, and management should decide on the respiratory protection required in individual circumstances on the basis of a formal risk assessment required under The Personal Protective Equipment Regulations 1992.
Respirators may take the form of:

- general purpose dust respirators;
- positive pressure powered dust respirators;
- helmet-contained positive pressure respirators;
- gas respirators:
    - cartridge type;
    - canister type; and
- emergency escape respirators.

RISK Risk expresses the likelihood or probability that the harm from a particular hazard will be realised.

RISK ASSESSMENT A risk assessment should usually involve identifying the hazards present in an undertaking (whether arising from work activities or from other factors, e.g. the layout of the premises) and then evaluating the risks involved, taking into account whatever precautions are being taken.
A suitable and sufficient risk assessment:

- should identify the significant risks arising out of the work;
- should enable the employer to identify and prioritize the measures that he needs to take to comply with the relevant statutory provisions; and
- should be appropriate to the nature of the work and such that it remains valid for a reasonable period of time.

(ACOP, The Management of Health and Safety at Work Regulations 1999.)

RISK RATING A risk rating is used as part of the risk assessment process and, in most cases, is the numerical product of probability (likelihood) and the severity of the outcome in terms of injury, damage or loss. Probability and severity may be measured on a scale.

SAFE SYSTEM OF WORK The integration of people, machinery and equipment in a correct environment to provide the safest possible conditions in a work situation.

SAFETY AUDIT The systematic measurement by an organisation's management of its health and safety programme against a series of specific standards. A safety audit subjects each area of an organisation's activities to a systematic critical examination.

SAFETY CULTURE The development of the right safety culture in an organisation is an important feature of the health and safety management process. This entails:

- acceptance of responsibility at and from the top, accompanied by the appropriate leadership and commitment;
- conviction that high standards are achievable;
- the setting and monitoring of relevant objectives/targets;
- systematic identification and assessment of hazards, installation and regular review of protective measures;
- immediate rectification of deficiencies;
- promotion and reward of enthusiasm and good results;

- acceptance that it is a long-term strategy that requires sustained effort and interest; and
- ownership at all levels of the organisation.

SAFETY DEVICE

A protective appliance, other than a guard, that eliminates or reduces danger before access to a danger point or area can be achieved.
Safety devices can take the form of trip devices, pressure sensitive mats, ultrasonic devices, two-hand control devices, overrun devices and mechanical restraint devices.
(**See TRIP DEVICE, TWO HAND CONTROL DEVICE, DEADSTOP DEVICE.**)

SAFETY INSPECTION

A safety inspection is generally taken to mean a scheduled inspection of a workplace or premises, to examine current levels of safety, working practices and compliance with the relevant statutory provisions and local systems at a particular time.
Safety inspections can be undertaken by managers, health and safety specialists, safety representatives and enforcement officers.

SAFETY REPRESENTATIVE

An employee appointed by his trade union to represent the members of that trade union in consultation with his employer on all matters affecting health and safety at work.
Legal requirements relating to safety representatives and safety committees are covered in The Safety Representatives and Safety Committees Regulations 1977.

SAFETY SIGN

- Prohibition signs – circular with a red band enclosing a crossed out symbol on a white background, such as 'NO SMOKING'.
- Warning signs – triangular in shape with a yellow background and black borders, symbols and text, such as 'LPG – HIGHLY FLAMMABLE'.
- Mandatory signs – rectangular in shape, incorporating a white mandatory symbol and/or text on a blue background, such as 'WEAR EYE PROTECTION'.
- Safe Condition signs – a green square or rectangle with symbols or lettering in white, such as 'FIRE EXIT'.

Under the Health and Safety (Safety Signs and Signals) Regulations 1996 employers must use a safety sign wherever there is a risk to health and safety that cannot be avoided or properly controlled by other means.
Signs must contain a pictogram instead of relying solely on text. Fire safety signs and signals are included and FIRE EXIT signs must incorporate the 'Running Man' pictogram.
Pipework containing dangerous substances must be marked, by fixing labels or signs at sampling and discharge points, together with small stores containing dangerous substances.

STATEMENT OF HEALTH AND SAFETY POLICY

There is a general duty under the HSWA for an employer to prepare and, as often as may be necessary, revise a written statement of his general policy with regard to health and safety at work of his employees and the organisation and arrangements for the time being in force for carrying out that policy, and to bring the statement and any revision of it to the notice of his employees.
The principal features of a statement of health and safety policy are:

- a statement of intent;
- the organisation, in terms of accountability and responsibility; and

- the arrangements that detail the systems and procedures for monitoring performance and overall implementation of the objectives detailed in the statement of intent.

STRESS

Stress can be defined in a number of ways:
- the common response to attack (Selye, 1936);
- a psychological response that follows failure to cope with problems;
- a feeling of sustained anxiety that, over a period of time, leads to disease; and
- the non-specific response of the body to any demands made upon it.

SUBSTANCE HAZARDOUS TO HEALTH

This means any substance (including any preparation) that is:
- listed in Part 1 of the approved supply list as dangerous for supply within the meaning of The Chemicals (Hazard Information and Packaging for Supply) (CHIP) Regulations and for which an indication of danger specified for the substance in Part V of that list is very toxic, toxic, harmful, corrosive or irritant;
- specified in Schedule 1 of the COSHH regulations (which lists substances assigned maximum exposure limits) or for which the HSC has approved an occupational exposure standard;
- a biological agent;
- dust of any kind when present at a substantial concentration in air; and
- not mentioned above, which creates a hazard to the health of any person that is comparable with the hazards created by substances the above-mentioned.

  (The Control of Substances Hazardous to Health Regulations 2002.)

'TOXIC' CLASSIFICATION

Substances and preparations that in low quantities can cause death or acute or chronic damage to health when inhaled, swallowed or absorbed via the skin.

TRAINING

The systematic development of attitude, knowledge and skill patterns required by an individual to perform adequately a given task.
(Department of Employment and Productivity (1978) Glossary of Training Terms: HMSO London.)

TRAINING NEED

A training need exists when the optimum solution to an organisation's problem is through some form of training.

TRIP DEVICE

A means whereby any approach by a person beyond the safe limit of working machinery causes the device to actuate and stop the machinery or reverse its motion.

TWO-HAND CONTROL DEVICE

A device that requires both hands to operate the machinery controls, thus affording a measure of protection from danger only to the machinery operator and not other persons.

USE (DISPLAY SCREEN EQUIPMENT)

This means use for or in connection with work.
(The Health and Safety (Display Screen Equipment) Regulations 1992.)

USER (DISPLAY SCREEN EQUIPMENT) An employee who habitually uses display screen equipment as a significant part of his normal work.
(The Health and Safety (Display Screen Equipment) Regulations 1992.)

VAPORISING LIQUID EXTINGUISHER This type of fire extinguisher consists of a cylinder containing the liquid that is pressurised with dry carbon dioxide or nitrogen. Striking a knob allows the pressure to expel the liquid.
Vaporising liquid extinguishers should not be used in confined spaces as the liquids and their combustion products are toxic.

VENTILATION Ventilation implies the movement of air through a building. This may be by natural or mechanical means or by a combination of both.
The principal features of a ventilation system are:

- the provision and maintenance of the circulation of fresh air in every occupied part of a workplace (comfort ventilation); and
- the rendering harmless of all potentially injurious airborne contaminants, e.g. dusts, fumes, gases (exhaust ventilation).

'VERY TOXIC' CLASSIFICATION Substances and preparations that in very low quantities can cause death or acute or chronic damage to health when inhaled, swallowed or absorbed via the skin.

VISUAL FATIGUE Eye strain and discomfort experienced by many people using display screen equipment, which may be due to a combination of both eyesight defects and poor workstation design. Vision screening of display screen equipment users should be undertaken on a regular basis and where users are experiencing the symptoms of visual fatigue.

WARNING SIGNS **(See SAFETY SIGN)**

WATER-CONTAINING EXTINGUISHER There are several types of water-containing extinguisher.

WORK EQUIPMENT Under The Provision and Use of Work Equipment Regulations 1998, 'work equipment' is defined as including any machinery, appliance, apparatus or tool and any assembly of components that, in order to achieve a common end, are arranged and controlled so that they function as a whole.

WORK-RELATED UPPER LIMB DISORDER A group of conditions occasionally experienced by keyboard operators, formerly referred to as repetitive strain injury. These disorders affect the soft tissues of the hand, wrist, arm and shoulder. Typical symptoms include pain, restriction of joint movement and swelling of the soft tissues.

WORKSTATION This means an assembly comprising:

- display screen equipment (whether provided with software determining the interface between the equipment and its operator or user, a keyboard or any other input device);
- any optional accessories to the display screen equipment;
- any disk drive, telephone, modem, printer, document holder, work chair, work desk, work surface or other item peripheral to the display screen equipment; and

- the immediate environment around the display screen equipment. (The Health and Safety (Display Screen Equipment) Regulations 1992.)

Abbreviations

Commonly used abbreviations in Intermediate Health & Safety

ACOP	Approved Code of Practice
CDM	Construction (Design and Management)
CHIP	Chemicals (Hazard Information and Packaging for Supply)
CIEH	Chartered Institute of Environmental Health
COSHH	Control of Substances Hazardous to Health
dB	decibel
DSE	display screen equipment
FPWR	Fire Precautions (Workplace) Regulations
HSE	Health & Safety Executive
HSWA	Health & Safety at Work Act
LEV	local exhaust ventilation
LPG	liquified petroleum gas
LTEL	long-term exposure limit
MEL	maximum exposure limit
mg/m^3	milligrams per cubic metre
MHSWR	Management of Health & Safety at Work Regulations
OEL	occupational exposure limit
OES	occupational exposure standard
PAT	portable appliance testing
PPE	personal protective equipment
ppm	parts per million
PUWER	Provision and Use of Work Equipment Regulations
RCD	residual current device
RIDDOR	Reporting of Injuries, Diseases and Dangerous Occurrences Regulations
RPE	respiratory protective equipment
RSI	repetitive strain injury
SRSCR	Safety Representatives and Safety Committees Regulations
STEL	short-term exposure limit
WHSWR	Workplace (Health, Safety and Welfare) Regulations

Bibliography & further reading

Management of Health and Safety at Work Regulations 1999, Approved Code of Practice and Guidance. HSE Books

Health and Safety Executive [1990]: *A Guide to the Health and Safety at Work etc. Act 1974. Guidance on the Act.* HMSO, London

Health and Safety at Work etc. Act 1974. HMSO, London

British Standards Institution [1996]: *Guide to Occupational Health and Safety Management systems (BS 8800).* BSI, London

Health and Safety Executive [1996]. *A Guide to the Health and Safety (Consultation with Employees) Regulations 1996.* HSE Books

Consulting Employees on Health and Safety. HSE Books

Health and Safety Executive [1991]: *Successful Health and Safety Management.* HMSO, London

Rimington, JR [1989]: *The Onshore Safety Regime:* HSE Director General's submission to the Piper Alpha Inquiry, December 1989

Stranks, J [2001]: *A Manager's Guide to Health and Safety at Work* (fifth edition). Kogan Page, London

Stranks, J [1994]: *Management Systems for Safety.* Pitman Publishing, London

Stranks, J [2000]: *The Handbook of Health and Safety Practice* (fifth edition). Pearson Education Limited, London

Index